# Migration, Ethics & Power

## SOCIETY AND SPACE SERIES

The *Society and Space* series explores the fascinating relationship between the spatial and the social. Each title draws on a range of modern and historical theories to offer important insights into the key cultural and political topics of our times, including migration, globalisation, race, gender, sexuality and technology. These stimulating and provocative books combine high intellectual standards with contemporary appeal for students of politics, international relations, sociology, philosophy, and human geography.

# Migration, Ethics & Power

## Spaces of Hospitality in International Politics

Dan Bulley

Los Angeles | London | New Delhi
Singapore | Washington DC | Melbourne

Los Angeles | London | New Delhi
Singapore | Washington DC | Melbourne

SAGE Publications Ltd
1 Oliver's Yard
55 City Road
London EC1Y 1SP

SAGE Publications Inc.
2455 Teller Road
Thousand Oaks, California 91320

SAGE Publications India Pvt Ltd
B 1/I 1 Mohan Cooperative Industrial Area
Mathura Road
New Delhi 110 044

SAGE Publications Asia-Pacific Pte Ltd
3 Church Street
#10-04 Samsung Hub
Singapore 049483

Editor: Robert Rojek
Editorial assistant: Matthew Oldfield
Production editor: Katherine Haw
Copyeditor: Solveig Gardner Servian
Proofreader: Neil Dowden
Indexer: Silvia Benvenuto
Marketing manager: Sally Ransom
Cover design: Wendy Scott
Typeset by: C&M Digitals (P) Ltd, Chennai, India
Printed and bound by CPI Group (UK) Ltd,
Croydon, CR0 4YY



First published 2017



**Library of Congress Control Number: 2016938531**

**British Library Cataloguing in Publication data**

A catalogue record for this book is available from the British Library

ISBN 978-1-4739-8502-5
ISBN 978-1-4739-8503-2 (pbk)

# Praise for *Migration, Ethics and Power*

'We live in an age of humanitarianism, a time when the highest form of politics and the noblest form of citizenship is to help others. This book places this new pastoralism in a different light. What if all this saving, sheltering and caring is not a new softness tempering the harsh edges of power but increasingly one of power's most privileged mechanisms? Richly contextualized, incisive and provocative, this book will change the way we understand hospitality. To the history of social struggle it adds a new antagonism: host/guest, truly a dialectic for our time.'

**William Walters, Professor of Political Science, Carleton University**

'*Migration, Ethics and Power* investigates the interaction of ethics and power in a range of spaces of hospitality that operate at and beyond the margins of the international state system. It astutely reveals that when people seek to cross borders or create spaces of their own, migration develops and hospitality is accessed, presumed, appropriated, or rebuffed. Dan Bulley's critical exploration and insightful analysis shows that while hospitality can be experienced as a right, a form of exchange, an act of compassion, or clandestine or abusive, it is always a matter of ethics, power and space. In this book, the everyday ethics of hospitality takes the form of diverse spaces – including homelands, refugee camps, global cities, postcolonial states, and a supranational community, from which the author uniquely explores the themes of post-sovereign ethics and power in international politics. Brilliantly perceptive and passionately engaged, *Migration, Ethics and Power* will be of great interest to those who are concerned with the everyday practices of hospitality, post-sovereign spaces, and international ethics and politics.'

**Suzan Ilcan, Professor of Sociology, University of Waterloo**

'Dan Bulley's *Migration, Ethics and Power* is a highly original contribution to the ethics and politics of international hospitality. It is philosophically astute, conceptually innovative and is distinguished by the depth of its political and ethical attunement.'

**Michael Shapiro, Professor of Political Science, University of Hawaii**

'Bulley's *Migration, Ethics and Power* is a timely reminder that beyond the oft-discussed statist politics of the refugee crisis there is an ethical response to the refugee crisis that is already unfolding within, and beyond, states. That hospitality is productive of specific subjectivities and spaces, from which more hopeful and humane relations might form. This is the right book for the right moment.'

**Jason Dittmer, Professor of Geography, University College London**

'This book is an untimely if urgent invitation to think about migration as an ethical and political subject through the paradoxes of hospitality/hostility. With a persistent focus on spaces where such paradoxes are played out – ranging from mosques, camps, cities, borders, hotels, homes, and states – it provides a challenging perspective on sovereignty as spatial practice across frontiers. Both students and researchers in migration studies, refugee studies, international studies, and citizenship studies will immensely benefit from engaging with this invitation to think differently about *where* sovereignty is practiced and **how** it functions.'

**Engin Isin, Professor of Politics, The Open University**

'This book provides a provocative discussion of hospitality as a key but often overlooked topic structuring the field of International Relations. From the refugee camp to the global city, this exploration provides a timely look into how various spaces are produced in relation to hospitality and the power relations within that both enable and limit its practice. Defining the "the crossing of borders" as the "hallmark of hospitality", this book will be an important read for those wishing to think seriously about the ethics of hospitality in light of growing numbers of people on the move but also out of place.'

**Kim Rygiel, Balsillie School of International Affairs, Wilfrid Laurier University**

# Contents

# About the Author

Dan Bulley is a Senior Lecturer in International Relations at Queen's University Belfast. His primary research interest is the way ethics and power become entangled in modern practices and spaces of international politics. Following earlier work on the ethics of British and EU foreign policy, his recent focus has been on the concept of 'hospitality' and its importance for understanding issues such as international ethics, immigration and asylum policy, EU enlargement and humanitarian intervention. He is particularly interested in how hospitality produces irruptive international spaces which might reveal new or unthought possibilities for our inevitably political relations with strangers.

# Acknowledgements

As always, a big debt of gratitude is owed to my wife, Bali. She is the only person to have read the book in its entirety before its delivery to SAGE, and has been a constant support throughout its writing, even whilst finishing her own book. Her comments were always useful, just as her encouragement was always well-meant ('at least you tried'), if not always helpful. Much of my engagement with Foucault in this book was prompted by reading drafts of her excellent work.

The support of many other people has been crucial in helping me finish this project. In particular, the listening ear of my mum, Carol Shergold, has been priceless. She's amazing. Bruno and Bella kept me sane and amused while my inestimable squash partners in Belfast – Brice Dickson, Marek Martyniszyn, Neil Matthews and Fabian Schuppert – allowed me the essential opportunity of hitting something really hard and swearing loudly. They have no idea how important this was.

The School of Politics, International Studies and Philosophy at Queen's University Belfast has been a very supportive environment in which to work on this book, and I will mourn its forthcoming merger. The sabbatical granted to me in the second half of 2015 has been invaluable, as has been the support of colleagues. In particular, Debbie Lisle has been a vital collaborator in developing some of the ideas in this book. Beverley Milton-Edwards offered support and advice on Chapter 4, whilst David Phinnemore's meticulously detailed comments on Chapter 5 were an inspiration (sadly, the specifics of Malta's exemptions from internal market rules did not make it into the final draft).

I would also like to thank friends and colleagues beyond Belfast who have helped me tighten, refine or completely change my arguments with their comments in conversations and engagement with different sections of the book in the form of conference papers. Gideon Baker, James Brassett, Jimmy Casas Klausen, Jonny Darling, Tarik Kochi, Tom Lundborg, Paddy McQueen, Vjosa Musliu, Cian O'Driscoll, Vicki Squire and Nick Vaughan-Williams have helped make this a much better book.

Finally, thanks to the editors at SAGE (particularly Robert Rojek and Matt Oldfield) and especially the *Society and Space* series editor, Stuart Elden. Stuart's interest in the book gave me great belief when I needed it most.

# List of Abbreviations

| | |
|---|---|
| AFAD | Turkey's Disaster and Emergency Management Presidency |
| AFSJ | Area of Freedom Security and Justice |
| BBC | British Broadcasting Corporation |
| BNP | British National Party |
| CARE | Cooperative for Assistance and Relief Everywhere |
| CAZ | London's Central Activities Zone |
| CEAS | Common European Asylum System |
| CEE | Central and Eastern European |
| EAM | European Agenda on Migration |
| EEC | European Economic Community |
| ESPON | European Spatial Planning Observation Network |
| EU | European Union |
| Eurodac | European Dactyloscopy |
| Eurosur | European Border Surveillance System |
| Frontex | European Agency for the Management of Operational Cooperation at the External Borders of the Member States of the European Union |
| GAMM | Global Agenda on Migration and Mobility |
| GDP | gross domestic product |
| GLA | Greater London Authority |
| ICTY | International Criminal Tribunal for the Former Yugoslavia |
| IDPs | internally displaced persons |
| INS | Immigration and Naturalization Service |
| IOM | International Organization for Migration |
| IR | international relations |
| ITN | Independent Television News |
| LDA | London Development Agency |
| LGBT | lesbian, gay, bisexual and transgender |
| MoL | Mayor of London |
| MoTA | Jordanian Ministry of Tourism and Antiquities |
| NATO | North Atlantic Treaty Organization |
| NGO | non-governmental organsation |
| OHR | Office of the High Representative (for Bosnia and Herzegovina) |
| PLO | Palestinian Liberation Organization |

| | |
|---|---|
| QUANGO | quasi-autonomous non-governmental organisation |
| RPPs | Regional Protection Programmes |
| RS | Republika Sprska |
| SAAs | Stabilisation and Association Agreements |
| TEU | Treaty on European Union |
| TFEU | Treaty on the Functioning of the European Union |
| TS | Transitional Settlement |
| UK | United Kingdom |
| UN | United Nations |
| UNHCR | United Nations High Commissioner for Refugees |
| UNICEF | United Nations Children's Emergency Fund |
| UNOPS | United Nations Office for Project Services |
| UNRWA | United Nations Relief and Works Agency |
| US | United States of America |
| USAID | United States Agency of International Development |
| USP | unique selling point |
| WFP | World Food Programme |

# Introduction

'Space is political and ideological. It is a product literally filled with ideologies.'

Henri Lefebvre

'All geographies are, in the last analysis, moral geographies.'

Michael J. Shapiro

In 2014 the Syrian conflict brought the ethics and politics of migration into stark relief. After three years of fighting that had, by that point, produced 2.5 million refugees and displaced 6.5 million people within Syria's non-functioning borders, the UN called on Western industrialised states to share the burden of hospitality through a fixed-quota system. By the end of the year Germany had welcomed around 20,000 and Sweden 9,000, whilst the UK government negotiated an opt-out from the deal in January 2014. Instead, it offered resettlement to 500 of the 'most vulnerable', including women and girls who had suffered sexual violence, the elderly, the disabled and victims of torture. In announcing the deal, Deputy Prime Minister Nick Clegg (2014) observed, with staggering cant:

> The £600 million we have provided makes us the second largest bilateral donor of humanitarian aid in the world. But as the conflict continues to force millions of Syrians from their homes, we need to make sure we are doing everything we can. We are one of the most open hearted countries in the world and I believe we have a moral responsibility to help ... Britain has a long and proud tradition of providing refuge at times of crisis. This coalition government will ensure it lives on.

Perhaps more surprising was the praise the UK government received from Maurice Wren, the Refugee Council Chief Executive who had been campaigning for greater openness to Syrian refugees since July 2013. In a press conference he suggested the UK was finally 'lead[ing] the way in offering refuge to people in their greatest hour of need'. He only hoped that 'other countries might now follow the UK's lead' in 'going the extra mile' (Refugee Council, 2014). By the end of 2014, the UK had welcomed 90 of these 'most vulnerable' refugees.

The UK government's policy and its representation are interesting for at least two reasons beyond their brazenness. First, in announcing a policy that appears overtly hostile in opting out of the UN's programme (itself not unproblematic), the morality of hospitality is both accepted and emphasised. This is significant because hospitality is recognised as self-evidently ethical, despite rarely figuring in accounts of international ethical behaviour in academia. Though included as a chapter in one recent textbook on international ethics (Shapcott, 2010: 87–121), this remains rare. More representative texts exclude hospitality entirely (e.g. Brown, 2002; Bell, 2010; Amstutz, 2013), focusing on traditional issues such as human rights, distributive justice, international law, the ethics of war, citizenship and terrorism. Likewise, most 'major' monographs in the field make no explicit allusion to hospitality. To take one example, liberal cosmopolitan Simon Caney's *Justice Beyond Borders: A Global Political Theory* (2005) discusses the usual topics of humanitarian intervention, just war and global distributive justice, but makes nothing of hospitality. 'Beyond' borders involves a one-way movement: global ethics and justice is limited to a particular type of border crossing, where those with plenty (whether of wealth, values, stable political structures or responsible military capacity) give from their abundance to those that lack.

The second point to be drawn from the fact that a state can practise such open contradiction without discomfort, and indeed be praised for it by a refugee advocacy non-governmental organisation (NGO), is that it underlines the exclusionary character of the modern, Western state. Andrew Linklater (1998: 147) has argued that the 'sovereign state is one of the main pillars of exclusion in the modern world'. Exclusion is a necessary hallmark of the Westphalian state and, though the same can be said of any space of hospitality, it remains overdetermined in nation-statist imaginaries and fantasies (Shapiro, 1997; Doty, 2003). What is profoundly restrictive, however, is how the state has come to form the limit of our understanding of international ethical practice and the horizon of its possibility – 'the state-oriented map continues to supply the moral geography that dominates what is ethically relevant' (Shapiro, 1994: 495). Most studies of international hospitality maintain the state as the central space and agent of welcome (Rosello, 2001; Cavallar, 2002; Benhabib, 2004; Brown, 2010; Baker, 2011). This state is all too often considered the primary actor in international ethics, the sole subject of responsibility to which everything else recurs (Jabri, 1998). Looked at in such a way, the possibilities of hospitality are profoundly disheartening. Though the UK may seem an extreme example, even Germany's largesse pales in comparison to a postcolonial state such as neighbouring Jordan, over half of whose population were refugees before welcoming over 600,000 Syrians (UNHCR, 2014a: 5–6). And any numerical approach to hospitality ignores the numerous exclusions and exercises of state power that emerge once the threshold has been crossed and refugees are placed in temporary camps or processes of 'resettlement'.

To think the politics and possibilities of international ethics we must therefore look beyond the 'statist imaginary' (Neocleous, 2003: 6). The imperative driving this book is the move towards what Mike Shapiro calls an 'ethics of post-sovereignty' by exploring the interaction of ethics and power in a range of spaces of hospitality that operate at the margins of the international state system. Instead of assuming the centrality of the state, an 'ethics of post-sovereignty must address the changing identity spaces that constitute the contemporary, unstable global map' (Shapiro, 1994: 488). Hospitality is an often-ignored practice that offers profound insight into these changing identity spaces. However, when compared to the spectacular focus of state-based international ethics which responds to genocide, ethnic cleansing, atrocities, natural disasters and acts of aggression through humanitarian interventions and just wars, hospitality appears less fascinating. After all, it is a more banal, 'everyday practice' (Still, 2010: 1). This may, in part, explain International Relations' (IR) traditional ignorance of hospitality. IR is a thoroughly masculine discipline which privileges the overtly violent and exceptional over mundane and everyday ethics (Onuf, 1998).

Yet hospitality's very 'everydayness' is important because migration is about far more than asylum granted by beneficent or hostile Western hosts. Wherever people seek to cross boundaries or sure up spaces they call their own, migration occurs and hospitality is offered, seized, assumed or refused. Whether those people are the wretched of the Earth (Chapters 1–2 and 5) or the global elite (Chapters 3 and 4), they depend upon hospitality and become one of two insecure, shifting subject-positions: host or guest. Guests can be refugees, but they can also be asylum seekers, regular and irregular migrants, travellers, traders, slaves, students, soldiers, sailors, labourers, tourists, terrorists, spies, diplomats, heads of state and government or professional athletes. All come to depend upon the generosity, cruelty or indifference of a host. The hospitality they experience may be a right or an economic transaction, it may be a form of exchange, an act of compassion or charity, or it may be clandestine and abusive. But it is always a matter of ethics, power and space.

Too often when it is touched upon, hospitality is assumed to be a straightforwardly ethical principle, an obvious 'good'.[1] We can see this in Clegg's claims about the UK government's policy toward Syrian refugees. Such treatment 'invites critical reflection' in order to explore whether hospitality 'conceal[s] an oppressive aspect beneath its welcoming surface' (Dikec, 2002: 228). Critical reflection must attend to the power relations that enable and limit practices of hospitality, the constraints they seek to impose and the struggle for control involved. I will go on to argue that hospitality is, as Jacques Derrida claimed, not only a crucial ethical principle, but ethics *itself* (1999a: 50). Yet this does not remove its equally necessary hostility, violence, power and resistance. This is why Derrida (2000, 2002a) coins the term 'hostipitality', an awkward neologism highlighting the inseparability of hospitality and hostility. While it is common to hear politicians in Western democracies complaining about migrants 'abusing our hospitality', perhaps it is better to see abuse as *constitutive* of hospitality as a practice, rather than something separate that can be done *to* it. Hosts and guests alike seek to abuse the generosity and vulnerability of their counterpart. The interaction of ethics and power is irreducible in a critical exploration of hospitality.

The aim of exploring an ethics of post-sovereignty is not, however, to produce a theory of post-sovereign international ethics; rather, it is to examine the 'habitus of experience' where ethical relations are not exceptional but always *ongoing*, where they are taking place and always entangled in 'spatial practices' (Campbell and Shapiro, 1999b: xii) involving tactics and technologies of power. This means not prejudging who is constituted as an ethical subject, who is host and who is guest, but always searching out 'subjects who are unfinished, ambiguously located, and enigmatic so as to resist the restriction of moral spaces to a state-oriented geographic imaginary' (1999b: xvii). These subjects are always becoming and never complete; guests becoming hosts and hosts becoming guests. They exercise power over themselves and others whilst resisting the management of their behaviour by the other, producing dynamic, shifting spaces of belonging and exclusion.

Within this overarching theme of the book, three central arguments emerge. First, I maintain that hospitality needs to be taken seriously in the interdisciplinary study of international ethics as a primary way in which 'we' practise everyday relations with difference. Hospitality is a central process by which 'we' negotiate and construct the identity of this 'we', its security, obligations and responsibilities. Though often ignored in academic and policy discussions of ethics, hospitality is perhaps the most common means by which popular culture imagines an ethical response to the worst excesses of politics: the violence of genocide and ethnic cleansing (see Chapter 1). Yet, in order to be taken seriously, we need a clearer understanding of hospitality, something Derrida warns us is impossible. Though he appears sure that ethics *is* hospitality, he also tells us we do not and cannot know what hospitality is (Derrida, 2000: 6). Much of the remainder of the Introduction will examine this aporia of definition.

My second central argument is that practices of hospitality involve not only the construction of ethical subjects (hosts and guests) and their relation (identity/difference, welcome/refusal, safety/threat), but also the production of spaces. This argument has been made elsewhere in relation to commercial tourist strategies that generate cities, hotels, cafes and leisure zones as more or less welcoming (see Bell, 2007a, 2007b; Lynch et al., 2011). Here, the case is made more broadly that hospitality is the means by which *particular* spaces are brought into being as 'homes', as embodying an *ethos*, a way of being: an *ethics*. Practices of hospitality carve out spaces as *mine* rather than yours, as places of belonging and non-belonging, and then manage and enforce their internal and external boundaries and behaviours. This concentration on space takes the focus of my argument away from those who stress the *temporal* aspects of hospitality, its moment of surprising encounter with the stranger as unexpected *arrivant* (Derrida and Dufourmantelle, 2000; Dikec, 2002; Barnett, 2005; Baker, 2011). Ruth Craggs

(2014: 90) argues that searching for such 'unanticipated and unplanned moments of welcome where a more open ethics of hospitality could be located' has meant overlooking 'welcomes that are planned, staged and expected'. While I examine elements of both (the unexpected always arises in the midst of meticulous planning), the expected arrival is a far more common way in which the everyday ethics of hospitality is practised. This allows me to focus on the space rather than the moment, exploring hospitality's generation of houses, hotels and homelands (Chapter 1), refugee camps (Chapter 2), global cities (Chapter 3), postcolonial states (Chapter 4) and a supranational community (Chapter 5). An almost infinite number of other spaces could have been chosen, including boats (Perera, 2002a), tents (McNulty, 2007: 35–41; Kaplan, 2011), international organisations (Craggs, 2014), churches, temples and mosques (Kearney and Taylor, 2011; Patsias and Williams, 2013). The spaces I choose are merely useful starting points for exploring the themes of post-sovereign ethics and power in international politics.

My third argument is that interrogating spaces of hospitality draws our attention to an aspect of international ethics that is often ignored: its inseparability from power. Any concrete practice of hospitality must *of necessity* involve relations of power and resistance between host and guest. These power relations both constitute its ethics and, in the process, contest and overturn the subject-positions of 'host' and 'guest'. After all, neither can ever be entirely sure of each other's identities, intentions or capabilities. They may be as they seem, but they may be lowering the others' defences, waiting to strike. So both host and guest undertake complex strategies and tactics of managing, evading, calculating, guiding, dodging, governing, resisting and surveilling. Yet, the moment a space is opened to the other – the first act of hospitality, perhaps – that space is destabilised and thrown into question, for good or ill, along with the subject-positions and power relations of host and guest. The benign host may be just that, but could also be abusive, violent or merely exploitative. They may themselves be a guest (a trafficker, perhaps) that cannot offer the promised welcome. A 'genuine' asylum seeker (guest) can become a settled refugee and, if they behave themselves properly, even a citizen (a host). Equally, they could be ejected to sure up the home. Likewise, the diplomat could be a spy, the tourist also a terrorist. Hospitality names this insecurity of ethics *and* the power relations that try to make it safe for both.

## HOSPITALITY: ETHICS, SPACE AND AFFECT

The importance of hospitality has recently been widely observed, with a remarkable rise in its popularity identified by scholars from a range of disciplines (Molz and Gibson, 2007; Dikec, Clark and Barnett, 2009; Still, 2011; Candea and da Col, 2012; Onuf, 2013). Despite this recognition, hospitality can be difficult to draw out or expand upon, in part because it is a practice we all know something about as we engage in or experience it on a daily basis. Yet it also has a long and venerable tradition in ethical thought, encompassing Greek antiquity (Bolchazy, 1993), the Natural Law tradition (Cavallar, 2002) and contemporary continental philosophy (Still, 2010). Tracy McNulty (2007: vii) even claims that the 'problem of hospitality is coextensive with the development of Western civilization'. For others, hospitality has a certain universality, in that it is considered an important practice across many cultures world-wide, and throughout large swathes of history. This 'fact' is often central to claims regarding its importance (Benhabib, 2004: 42–43; Friese, 2004: 71; Still, 2011: 1, 2007: 194; Candea, 2012: 46; Pitt-Rivers, 2012: 517). Yet upon closer inspection, the diverse and particular codes, rules, norms and laws of hospitality at the very least temper, if not explode, strong universality claims. It is hard to see the strategic, generous, 'ironic and irreverent' hospitality practised by mobile Afghan traders in the borderlands of Central Asia (see Marsden, 2012: 118–124) as part of the same universal practice as the 'satisfying medium' hospitality provided by middle-class white Americans (Dennis, 2008: 19), or Ancient Greek hospitality based on a fear of the gods and maintenance of elitist social structures (Bolchazy, 1993). Certainly they are not the same in any straightforward sense.

While perhaps universally important, hospitality's particular forms vary widely across time and space; it does not easily give itself to stable representation. It is tricky and hard to pin down. What does unite all forms, however, is that any practice of hospitality will inevitably be torn between complete openness and varying degrees of closure. This forms the divide between what Derrida calls *the* law of unconditional, unquestioning, absolutely open hospitality and the conditional, interrogating and restricting law*s* of hospitality (Derrida and Dufourmantelle, 2000: 77–91). The two are entirely distinct and yet cannot be separated; each requires the other yet they are violently opposed. While unconditional hospitality as an absolute openness is impossible to organise into policy (Derrida, 2003: 129, 1999a: 90), its openness would also destroy the host as such, their mastery of the home that makes hospitality possible in the first place (Derrida, 2002a: 364). In place of host and home we would have indeterminate being and space. Unconditional hospitality is not even *desirable* as it requires the host's destruction and offers nothing definite or material to the migrant, no security, shelter or sustenance, none of the 'goods' of hospitality. And yet, the conditional law*s* of hospitality offer no resolution; as conditions on an unconditional, they will always be annulments of *the* Law. By setting conditions these laws threaten to remove themselves from the domain of hospitality as such. What prevents them from becoming merely laws (as opposed to laws *of hospitality*) is that they retain at least the thought of the unconditional (Derrida, 2003: 129) and the fact that they can *never* achieve their aim: 'no regulation finally can master the exposure to the visitation of others' (Hagglund, 2008: 104).

The universality of hospitality is contained in the fact that all its practices need to negotiate and renegotiate between *the* law and the law*s*. Conversely, this irreducible negotiation is also at the heart of hospitality's variety and particularity. It is a reason to be wary of all attempts to tame the concept: it 'rebels against any self-identity, or any consistent, stable, and objectifiable conceptual determination' (Derrida, 2000: 6). Despite this warning, Derrida gives us the surprising assurance that hospitality *is* 'ethicity itself, the whole and the principle of ethics' (1999a: 50). This is not, as it might first appear, a call for ethics as unconditional hospitality. Such an absolute openness is both undesirable and impossible to organise. Rather, ethics is hospitality because every individual negotiation of openness and conditionality is an expression of an '*ethos*, our way of being, the residence, one's home, the familiar place of dwelling, the manner in which we relate to ourselves and to others, to others as our own or as foreigners' (Derrida, 2001: 16–17). Ethics is hospitality not because hospitality is what we *ought* to do, but rather because it is what we *do* do, in every moment, as a way of being in relation to ourselves and others. Every singular negotiation of *the* law and the law*s* is a particular practice of ethics, an expression of an ethos as a manner of being; it is always 'answering for a dwelling place, for one's identity, one's space, one's limits' (Derrida and Dufourmantelle, 2000: 149).

It is precisely these particular practices of ethics as hospitality that I want to examine in this book. Therefore, while heeding Derrida's warning that any general conceptual definition is impossible, an unstable and inconsistent attempt at determination is important if we are to consider hospitality as an ethics of post-sovereignty.[2] Such minimalist taming also wards off an expansionist approach in which hospitality comes to encompass any social relation or responsibility-taking.[3] Therefore, I want to argue that hospitality is a *spatial relational practice with affective dimensions*. It is this combination of the spatial and affective which makes hospitality a complex interplay of ethics and power relations.

Hospitality is a *spatial* practice because it requires the crossing of borders and thresholds (Still, 2010). It demands an inside and an outside, and that the two be separated by more or less clearly defined and understood boundaries. But for hospitality to occur there must be a breaching of the frontiers that divide the two; whether the borders exist in the form of the walls and doors of a house, the ephemeral and shifting boundaries separating communities, cities, states or regions, they must be crossed. That which belongs outside must move inside, whether it is called or arrives unannounced and uninvited. According to Doreen Massey (2005: 71), space is a sphere of coexistence containing a multiplicity of trajectories, involving previously unrelated subjects and objects, people and things, coming into contact with each other. In this sense, hospitality as a spatial practice is one of organising space, delimiting it, taming and freeing it as a sphere of coexistence, restricting and enabling its multiplicity, regulating, filtering and channelling trajectories and contacts that it allows; preventing the 'bad' trajectories, enabling the 'good', while missing many altogether. More than simply organising it, hospitality *generates* or *produces* a space; it is no longer just any indeterminate sphere of coexistence or any kind of relation or contact, it is determined and regulated.

It makes the space *your* space rather than *my* space, and the practice of my crossing into your space is conducted on this basis. Hospitality brings a space into being, cordoning it off from other spaces, as *this* rather than *that* – yours rather than mine, private rather than public, individual rather than communal, home rather than away, domestic rather than international, ordered rather than anarchic. The outside constitutes the inside, but never completely: hospitality ensures that the outside is already within, that the space is never established as an impregnable fortress but through a permeability that welcomes, rejects and can be surprised by the unexpected and unseen arrival.

This means that, while hospitality presupposes the possibility of strictly applied frontiers that separate spaces of identity and difference, self and other (Derrida, 1999a: 92; Derrida and Dufourmantelle, 2000: 47–49), it also requires that these frontiers open, allowing crossing. The work of hospitality then is one of both preserving *and* disturbing this inside as a carefully delimited and controlled space, by allowing the outside in. It is bolstering and disruptive, consolidating and transformative (Selwyn, 2000: 19). As I will go on to show, the crossing of boundaries that is the hallmark of hospitality means that it creates a space which is both this *and* that, or this *becoming* that: it is the becoming public of the private, the often carefully circumscribed and managed disruption of the home with the away, the more or less internationalising of the domestic. It generates hybridity, leaving only nostalgia for homogeneity and the desire to escape undecidability (Doty, 2003: 12). Hospitality requires a settled spatial relation that it also upsets, its attempt to contain and govern this disruption thereby constructing a particular, contingent and contested space.

The spatial nature of hospitality also helps distinguish it from other forms of ethical practice. For example, giving money or food to a stranger on the street could be a generous or charitable gesture, but no spatial boundaries need be crossed. Likewise, stepping in to resolve a quarrel in the street or a bar could be a responsible act, but it is not hospitable because it takes place in a shared space. Humanitarian NGOs like Médecins Sans Frontières offer free medical care to victims of war or natural disaster, but their actions are not themselves hospitable (though they may require the hospitality of a state or region to undertake that work). When Human Rights Watch monitors governments and campaigns for victims of oppression, their work can be compassionate and caring but it is rarely hospitable. If a state or international organisation such as NATO intervenes in the affairs of a sovereign state that is committing atrocities, such an act could be just or unjust, responsible or irresponsible, but it is not hospitable. Rather, it often involves a forceful violation of another's space. Hospitality practices are more than just encounters; they are specific types of encounter requiring spatial boundaries that are simultaneously displaced through acts of crossing and dwelling.

This spatial understanding is certainly unfaithful to the temporal focus of hospitality as the moment of encounter contained in the works of Derrida and Emmanuel Levinas (for more faithful readings, see Barnett, 2005; Baker, 2011).

Concentrating on the spatial effectively extends the temporality of hospitality as a practice that continues *beyond* the threshold moment. The spatial thus remains a relatively under-explored element within Derrida's thought of ethics as hospitality. Yet focusing on space brings us closer to everyday practices and experiences of hospitality where what happens *after* the encounter is equally important. It allows a deeper analysis of the relation between ethics and power as co-existence within a shared space is negotiated. It is also in part this spatial aspect, along with its affective dimensions, that makes hospitality a particular kind of ethical practice. Hospitality does not occur in an *indeterminate* space and nor does it involve the traversing of *non-meaningful* boundaries. What gives the space and boundaries their determinacy and meaning is their affective dimensions; they constitute lines between feelings and emotions of belonging and non-belonging, comfort and discomfort, security and insecurity, ease and awkwardness. The space produced by hospitality is the *home*, along with its affective sense of being-at-home-with-oneself. This is what hospitality both enables and upsets.

A home is therefore a spatial *and affective* articulation of an ethos. While it is partially defined by its inside-outside structure, a home is just as much a matter of feelings and emotions (Blunt and Dowling, 2006: 22), whether they be of strangeness or familiarity, homeliness or unhomeliness. This fusion of the spatial and affective is what links the home to an identity, to a *manner* of individual or communal being and dwelling. Such an understanding of the home does not essentialise or depoliticise the concept. Treating the home as a product of affect is not to romanticise it as a safe or pure place of belonging, a static, 'comforting bounded enclosure' (Massey, 1992: 12). It can never be simply opposed to the 'away' of uncertainty, strangeness and danger (Honig, 1994: 585–586; Ahmed, 2000: 87–88). Rather, the emotional and structural boundaries of the home are both the continuing source and product of contestation, movement and violence (Massey, 1992: 14). The home is therefore never politically neutral. It is never the location of a stable, secure ethos, 'uncontaminated by movement, desire or difference' (Ahmed, 2000: 88). Rather, a home is always the product of selected inclusions and exclusions, a 'negotiated stance' (George, 1996: 2). This ethico-political negotiation is *hospitality*, the practice by which the subject and its ethos is produced and managed along with its boundaries and openings, its spatial and affective sense of belonging and non-belonging, identity and difference, satiety and desire.

This understanding of hospitality can be closely tied to what Blunt and Dowling (2006) call a 'critical geography of the home'.[4] They argue for an understanding of 'home' as a '*spatial imaginary*' (1996: 2), rather than something fixed and timeless; an imaginary through which we negotiate our ethos, identity and subjectivity. Conceived as such, we can no longer limit the home to a particular scale, that of the house or the state imagined as a 'homeland' (see Neocleous, 2003: 98–124). Rather, it is necessarily 'multi-scalar'. Imaginaries of home and home-making negotiations can occur at any level, from the family, workplace, community, suburb, neighbourhood, nation, homeland or region (Duyvendak, 2011: 111) to 'the park bench' (Blunt and Dowling, 1996: 29). Furthermore, homes are constituted through relationships between both humans and the non-human which can equally be invested with emotions, memories, familiarity and comfort (Miller, 2008). In this book, homes are examined as houses, hotels, homelands, refugee camps, cities, states and supranational communities. If a home is the affective and spatial imaginary produced through the ethico-political negotiations of inclusion and exclusion (i.e. hospitality), it cannot be restricted to one particular space.

None of this, however, suggests that hospitality is a practice performed by a stable subject (an 'I', 'we', 'one' or 'our') with a pre-existing spatial and affective home or ethos. Rather, the home and its ethos are *co-produced* with the migrant or stranger, that which does not belong. The 'host' and 'guest' as ethical subjects are only ever present as such through their encounters and negotiations. After all, the space, limits and boundaries of home and ethos only retain any meaning because of their opposite: spatial and affective *difference*. The granting or refusing of hospitality describes a particular coming-into-relation that (re)creates, sures up and transgresses the boundaries that produce and enable the home, the self and its ethos. This is further complicated by the fact that we cannot go back to an originary moment of contact, a first coming-into-relation. As Jean-Luc Nancy (2000: 29) has argued, 'we' have never been singular but have always been plurally singular and singularly plural. Subjects did not simply emerge as separate entities, but through an original exposure to each other; existence has only ever been co-existence (Nancy, 1991: 58). Hospitality and hostility are thereby practices of relation in which identity and difference, self and other, spaces of belonging and non-belonging, home and away have been violently carved from an undeniable ontological coexistence. The fact that these practices are so imbued into the everyday processes of international politics makes them crucially important for how we think and theorise an ethics of post-sovereignty; but it also makes them incredibly mundane, such that they can easily escape the attention of a discipline like IR which is so focused on the exceptional and extreme.

## MANAGING HOSPITALITY: ETHICS AND POWER

Explicit discussion of power is often avoided in explorations of international ethics. An important exception here is the Realist tradition, in which the pursuit of power in the national interest is the highest of moral duties (Morgenthau, 1951: 242; Donnelly, 2000: 164–167). This conflation of ethics and power is more often perceived as *amoral* by those holding to a universal conception of ethics. Even some of its supporters hold that such pure power politics means the 'preconditions for morality are absent' (Art and Waltz, 1983: 6). Beyond Realism, it is customary for cosmopolitan approaches to seek a transcendence of the power relations that seem to dominate the international (Pogge, 1989: 218–229), or a rolling back of the powers and monopoly on citizenship of the sovereign state (Linklater, 1998: 179–212). More common still is to avoid discussion of power altogether. Indeed, Molly Cochran (1999: 228) describes it as a 'formidable criticism', primarily emerging from Feminists, that the question of power has not been 'adequately addressed in normative IR theory'.

At the basis of this evasion is perhaps that, even when the thought is not made explicit, we retain a hankering for a conception of the ethical as the pure, the decidable and non-negotiable. Yet, retaining this view of ethics as beyond power makes it difficult to engage with everyday politics, its compromises and 'dirty hands' (Walzer, 1973). This is perhaps why Constructivists in IR have fastened on to 'norms' and avoided too much discussion of ethics and morality; their idea of the normative does not commit them to anything as pure and undiluted as 'ethics' (e.g. Thomas, 2001; Manners, 2002; for a critique, see Bulley, 2014b). Thus, power tends to be seen as the opposite of ethics, or else an ethical rider or prefix must be added to power to dampen and ameliorate its 'hard' edges, as in discussions of 'soft' power (Nye, 2004), 'civilian' power (Duchêne, 1972; Bull, 1982), or 'normative power' (Manners, 2002, 2006). Certainly power does not have an easy relationship with ethics; if they mix beyond Realism's conflation, ethics appears a slightly unnatural softening and civilising outside influence on power. But they do not belong together; one somehow diminishes the other.

However, the only way we can free ethics from power relations is through complete abstraction; the moment we discuss immanent or concrete practices, power impinges on all relations and encounters with difference. Practices of charity, aid, human rights promotion, humanitarianism and care often assume an abundance of goods on the part of the 'giver' (expertise, knowledge, strength, material capabilities or conscience). These are implicitly or explicitly associated with moral capacities that those being helped are seen to lack. As postcolonial and poststructural approaches have demonstrated (Doty, 1996; Mutua, 2001; Orford, 2003; Korf, 2007; Fassin, 2012), whether motivated by pity, compassion or responsibility, these practices erect a moral hierarchy of actors (from the base of those lacking, to the peak of those overflowing with abundance). This hierarchy is then used to exercise further power, justifying surveillance, channelling and controlling aid, intervention, or financial and political governance (Duffield, 2001; Barnett, 2011; Ilcan and Lacey, 2011). We cannot free these practices from power, but they can be examined contextually to draw out how power is operating and what it is *doing*, both to and through claims to compassion, human rights, humanitarianism and obligation.

Hospitality is particularly interesting in this regard because, perhaps unlike the giving of aid or acts of intervention, it can involve as much disruption of the self/host as the other/guest. The host with an apparent 'abundance' of shelter, food and resources can be made to feel as insecure and threatened as the guest by their presence at the door or within the home. Hospitality therefore involves at least a fleeting glimpse of equality, 'the cosmopolitanism of a moment' (Kristeva, 1991: 11). Unlike humanitarian aid or intervention, hospitality requires no movement of the abundant toward the lacking, the strong to the weak, but can be entirely passive or inadvertent on the part of the abundant, working in the blind spots and invisibilities generated by its strength (see Chapter 3). Hospitality can – and often does – work by subverting distinctions and rendering the weak strong and the strong weak (see Chapters 2 and 4).

The ways in which ethics and power interact in and through hospitality cannot be reached, however, by relying on Derrida's reading of the concept (Bulley, 2015). Throughout his writings, Derrida traces hospitality to the question of the state and the decision of the sovereign (Derrida and Dufourmantelle, 2000: 47–79; Derrida, 2001: 22–23, 1999a: 20–21, 2003: 127–129; Derrida and Stiegler, 2002: 17–19). He frequently refers back to the etymological research of Émile Benveniste, who ties hospitality to the Latin chain *hosti-pet*. Within this sequence, host and guest, *hostis* and *hospes*, links to *potis*, *potestas* and *ipse* (Benveniste, 1973: 71; see Derrida, 1998: 14 and fn. 2; Derrida and Dufourmantelle, 2000: 41). Linguistically, hospitality is thereby entangled with both power (*poti-*, *potior* – 'to have power over something, have something at one's disposal' (Benveniste, 1973: 74)) and self-identity (through *ipse*). The power exercised through hospitality is treated as a sovereign relation: a 'system of domination' (Foucault, 1998: 92) based on sovereign possession of the home and its space (Derrida and Dufourmantelle, 2000: 41). Traditionally, then, the power of hospitality is that of 'mastery' for Derrida (2000: 13), mastery of the self, the self's space and the self-same (*ipse*) (Derrida and Stiegler, 2002: 111). It is about controlling thresholds and the decision to welcome or reject (Derrida, 2000: 14). While Derrida works to deconstruct these customary understandings, his readings start and end with the power of the masculine sovereign to decide and control the borders of the home, often equated with the state.

Conventional IR has followed this defining structural differentiation and separation of inside (the state) from outside (the state), with the border between the two monitored and maintained through sovereign power (Ashley, 1987; Walker, 1993; Agnew, 2007; Vaughan-Williams, 2009). However, in traditional IR theory, this structuralism has led to the factoring out of ethical issues (Williams, 2003; Hutchings, 2007), questions of affect and belonging (Ehata, 2013), and the power relations and violence that characterise 'domestic' life (Sylvester, 1994; Enloe, 1996). If the state, the 'inside', is a realm of hierarchy and sovereignty, the 'outside' is anarchic with only residual and easily relinquished forms of society, community, ethics and justice. In contrast, for Derrida it is this sovereign power over the home that *allows* the equation of hospitality as ethics described earlier. Without mastery over the space and its *ethos* hospitality cannot operate or offer anything. But this equation does not lead to a simple affirmation of sovereign mastery; rather, it generates a series of aporias.

The first aporia arises from the fact that such enabling sovereign power also places a limit on that which cannot be limited, in its own terms:

> It does not seem to me that I am able to open up or offer hospitality, however generous, even in order to be generous, without reaffirming: this is mine, I am at home, you are welcome in my home, without any implication of 'make yourself at home' but on condition that you observe the rules of hospitality by respecting the being-at-home of my home, the being-itself of what I am. There is almost an axiom of self-limitation or self-contradiction in the law of hospitality. As a reaffirmation of mastery and being oneself in one's own home, from the outset hospitality limits itself at its very beginning, it remains forever on the threshold of itself. (Derrida, 2000: 14)

Self-mastery and sovereignty are what is preserved by conditional laws that filter and select; they are essential to hospitality and yet radically undermine it. Without sovereignty, 'we' would not be a 'we' and could offer nothing to the migrant by welcoming them in, and hospitality would be ineffective 'without, in some concrete way, giving *something determinate*' (2003: 129, 1999b: 69). But we choose and limit *what* and *how* we offer it. An unconditional offer of 'make yourself at home' would relinquish mastery, making the host the guest and the guest the host, or dissolving the distinction and thereby offering *nothing*. The potentially genocidal effects of such a welcome are illustrated by Gideon Baker (2011: 27–37), and I examine a milder form in Chapter 4.

A second aporia emerges from the fact that, no matter how conditional, *any* practice of hospitality undermines sovereign mastery. As soon as the stranger crosses into the home, the home is no longer the self-same over which the sovereign master has complete control. Being is no longer at-home-with-itself, but at-home-with-the-stranger. The home and its *ethos* are transformed, contested, made vulnerable and on the verge of destruction. 'Even the most conditional hospitality is unconditionally hospitable to that which may ruin it. When I open my door for someone else, I open myself to someone who can destroy my home or my life, regardless of what rules I try to enforce on him or her or it' (Hagglund, 2008: 104). Thus the undecidability that Derrida reveals simultaneously reverses the power of sovereign mastery and displaces it.[5] The distinction between sovereign host and powerless guest is dissolved. We are left with the *hôte*, that which in French means both 'host' and 'guest' while also indicating the impossibility of either as stable identities (Derrida, 1999a: 41–42).

Derrida's reading of hospitality rigorously uncovers the deconstruction of sovereign mastery. However, partly because of this concern, his overwhelming focus is on the moment of encounter and the decision of inclusion or exclusion; it is always a matter of thresholds. But what powers are exercised in practices of hospitality once the threshold is crossed? What happens when we extend the temporality of hospitality and deepen its spatiality? As Julia Kristeva (1991: 11) notes of the hospitable encounter, 'the meeting owes its success to its temporary nature, and it would be torn by conflicts if it were to be extended'. Once inside, both host and guest are destabilised and dislocated. Both may seek to re-establish a sovereign mastery, but they can only do so by *ending* the hospitable relation and casting the other out. Once the outside is inside, as hospitality dictates that it must be, how is power exercised to manage and control the space of the home and its affective relations of belonging and non-belonging?[6]

There are severe limitations to Derrida's treatment of hospitality as a 'scale free abstraction' (Candea, 2012) rather than a concrete practice. Once we turn to such specific enactions and individual constitutions of the home, as this book does, we need to ask different types of questions to Derrida's: How do the power relations differ depending upon the post-sovereign space being constituted as home in a *particular* practice of hospitality? How do they produce more or less inclusive exclusions and exclusive inclusions? What kind of subjects or identities arise when 'host' and 'guest' are displaced or 'torn' by their conflicts? Does the management of space, affect and subjects multiply and diffuse thresholds throughout the home, or has the threshold itself been indefinitely deferred? To closely follow Derrida's focus on the threshold moment of hospitality limits our investigations of its power and ethics to that of an ever-deconstructing sovereign gift, avoiding these questions. We can perhaps see this in the way his analysis has been used in IR. For instance, in critiquing the use of identity cards to control the movement of refugees, Peter Nyers (2006: 94) notes that, though initially 'promoted as a technology of hospitality', they 'almost immediately became yet another technology of control'. But why must hospitality and control be opposed? To gain a more thoroughgoing understanding of how post-sovereign international hospitalities work ethically and politically, in all their complexity, we need to go beyond a focus on sovereign decisions and mastery. We must seek out the technologies, tactics, norms, rules and laws of welcome and control, viewing them as inextricably tied together and constitutive *of* hospitality. To do so is to explore how hospitality is *managed* and *controlled*; indeed, how ethics itself is governed, secured, immunised and made less risky.

## STRUCTURE OF THE BOOK

The book proceeds by examining a range of practices of hospitality and their production of 'homes' on a number of different scales. Chapter 1 investigates 'Genocidal Hospitality', arguing that, while academia and states continue to side-line hospitality as an ethical practice in situations such as the Syrian war, popular culture appears to do the opposite. Cinema in particular privileges hospitality as the primary way in which we imagine the interpersonal moral response to the 'worst' forms of violence which forcibly displace people. This is demonstrated through a critique of three films from different conflicts:

Rwanda in 1994 (*Hotel Rwanda*); Bosnia from 1992–95 (*Welcome to Sarajevo*); and Armenia in 1915 (*Ararat*). Through a critical reading of these films, the chapter reveals the inter-relation between ethics as hospitality and the power relations that unsettle a simply ethical narrative. I provocatively suggest that the sovereign, pastoral and governmentalised hospitality offered by Western homes and homelands demands an erasure or disavowal of difference which makes it distinct from, but not unrelated to, the genocide from which it saves the stranger.

While popular culture privileges interpersonal hospitality, the vast number of people displaced by conflicts do not find shelter in Western homes; they remain in the global South. Chapter 2, 'Humanitarian Hospitality', examines the welcome that produces and governs refugee camps as diverse, temporary, humanitarian spaces for 'saving' such people. By inspecting the refugee camp's assembled host and its strategies for the construction and government of these camps, we see how humanitarian hospitality generates a particular kind of homely space. Through a variation of domopolitics, it tries to ensure a guest population's compliance through a negotiation of non-belonging. However, turning to anthropologies of refugee camps exposes the way camp spaces are being variously seized and redirected through counter-hospitalities that deconstruct and resist the identities imposed upon refugees. Resistance thus operates by generating different forms of homeliness and ethos which temporarily reverse the power relations between host and guest.

Chapter 3, 'Flourishing Hospitality', begins by noting that most Syrian refugees have not sought the hospitality of camps but rather that of cities. Global cities in particular are constituted through their openness to global flows of people, things and ideas. I draw out an urban ethos as one of freedom and flourishing which operates via tactics of security that promote some circulations and mobilities whilst restricting others. This has important implications for who can be an acceptable guest and who becomes a liminal guest-host, or (g)host – the figure that does much of the city's embodied hosting and yet is a nominal guest. Concentrating on the spatial planning of London, we find a highly planned urban 'home' that directs mobility and behaviour whilst intervening in every aspect of the infrastructure and environment. This hospitality is built on the flourishing of both host and guest, alongside the exploitation of the (g)host. Such a governmentalised hospitality is always being countered and redirected by (g)hosts and hosts, yet transnational transiency and sanctuary city movements alike ultimately work to bolster the urban ethos of government and welcome as well as resisting it.

Analysing international hospitality without discussing colonialism is deeply negligent (Rosello, 2001). After all, postcolonial territories were formed into states by the welcome forced upon them by strangers. Chapter 4, 'Unconditional Hospitality', centres on the postcolonial state of Jordan, which has offered an exceptional welcome to refugees fleeing the conflict in Syria. Instead of setting its practices in the context of past hospitality to refugees, I offer an alternative history, a brief and episodic genealogy of how Jordan has been materially, institutionally, commercially and ethically constructed through its apparently unconditionally hospitable encounter with guests. Where earlier chapters concentrate on 'strong' hosts and 'weak' guests, here the relationship is reversed. Its lack of formal sovereignty sets Jordan within a postcolonial frame before I draw out the productive clash of *ethea* between colonial hosts and their colonial guests. The resulting exercise of adaptive and patriarchal disciplinary control galvanised the borders of the emerging (Trans-)Jordan, pacifying the desert and forming a non-sovereign postcolonial national identity and ethos. In more recent times, this ethos has been further disciplined, instrumentalised and monetised for commercial gain with the help of new guests. Thus, when Jordan's ethics of unconditional hospitality towards refugees is praised, this alternative genealogy sets such praise within the play of dominations that has produced the contested borders and ethos of a pre-, post- or non-sovereign space.

The final chapter, '(Auto)Immunising Hospitality', turns to EUrope and its current hospitality crisis. As a space which identifies itself as a territorial home defined by its ethos, EUrope sometimes appears particularly welcoming of difference. I argue that EUrope's ethos was unstably 'founded' upon a confrontation with a potentially threatening outside. Its practices of hospitality – both to states through its enlargement and individuals via immigration and asylum policy – are an immunising response to that perceived menace. This operates, on the one hand, through a pastoral ethopolitics which guides neighbouring states towards the protective EUropean home, helping them become EUropean. On the other hand, a pastoral biopolitics protects and cares for refugees and migrants in spaces which, outside the EU, are *becoming* but will never completely *become* EUropean. Such immunising hospitality, however, necessarily fails due to the *auto*immune nature of the EUropean ethos – the resistance to itself caught within its 'founding' principles of liberty, democracy, human rights and solidarity. Thus, the crisis in which EUrope recently finds itself is not just one of hospitality, migration or refugees, but one which put its very 'self' into question.

The Conclusion asks what hope or possibilities have been opened up by its analysis against the background of the coordinated attacks in Paris on 13 November 2015. These events underline the risks of hospitality, the vulnerabilities necessary to its practice and the relations of power and resistance it encompasses. The understanding of hospitality as an ethics of post-sovereignty can offer no normative rules or policy proposals in the face of such risks. Instead, the Conclusion asks what resources and opportunities can be exposed by the relentless critique of post-sovereign spaces and the ethics they articulate through practices of hospitality.

## NOTES

1. A more nuanced approach is often taken by critically informed studies where hospitality is brought in, often at the end of analyses, as an ethics and politics that can problematically redeem operations of power exemplified by the camp (Diken and Laustsen, 2005: 184–188), international borders (Khosravi, 2010: 125–129; Vaughan-Williams, 2015: 144–147), Europe's multicultural identity politics (Amin, 2004: 14–17), or citizenship laws (Benhabib, 2004: 35–40).
2. Despite Derrida's warnings that we cannot know what hospitality is, he moves between abstract claims about hospitality – as an 'interruption of the self' (1999a: 51) – to discussions of hospitality as a concrete practice – especially with regard to the *sans-papiers* (2002b, 2005a). In such moments, he is working with some kind of determination, though his reasoning is not made explicit.
3. For instance, in Gideon Baker's Levinasian reading of humanitarian intervention he claims it cannot be 'finally separated' from hospitality (2011: 111). This argument relies upon a move from hospitality read as an abstraction (where it is inextricably linked to justice, responsibility and so on) to a concrete practice, where hospitality and humanitarian intervention seem necessarily distinct. We can see this when he observes that the host 'goes forth' beyond its own borders when engaging in humanitarian intervention. This is to confuse a 'host' with an 'agent'; when the host leaves its home, it is no longer a host but an often unwanted guest.
4. For an excellent overview of the literature on critical geographies of the home, see Brickell (2012).
5. Michael Naas (2008: 62–80) and Martin Hagglund (2008: 20–30, 180–191) argue convincingly that the deconstruction of sovereignty as a theologico-political concept has been a key theme running throughout Derrida's work.
6. Derrida is not unaware of the diverse power relations operating once the threshold is crossed. After claiming that '*ethics is hospitality*' he observes it 'supposes a reception or inclusion of the other which one seeks to appropriate, control, and master according to different modalities of violence' (Derrida, 2001: 17). Focusing on reception and inclusion however, he does not expand on the modalities of violence and power such efforts to control produce.

# 1

# Genocidal Hospitality: Homes, Hotels and Homelands

Whilst it is important to stress the quotidian nature of hospitality as an ethical practice, this does not limit it to the banal and mundane. In fact, when faced with perhaps the most appalling emergency, hospitality often appears the obvious ethical action on an interpersonal level. Perhaps the best-known tales from the Holocaust, those of Anne Frank and Oscar Schindler, are based in acts of hospitality which at least temporarily saved lives. The salvational potential of hospitality thus appears obvious. Because of the nature of the horrific acts that make up genocide, crimes against humanity and ethnic cleansing (from enforced displacement to mass murder and rape), and the way they target a specific identity or ethos within a population, the act of hiding or protecting a representative of such a group in your own space seems an uncomplex if risky ethical act. Because of this, non-documentary films about genocide frequently turn upon moments of hospitality, as I will show in the three films which form the focus of this chapter: *Welcome to Sarajevo* (1997), *Ararat* (2002) and *Hotel Rwanda* (2004). But it is a trope shared by many others, including *The Mission* (1986), *Schindler's List* (1993), *Rabbit-Proof Fence* (2002), *Shooting Dogs* (2005), *Darfur* (2009), *The Boy in the Striped Pyjamas* (2011) and *Sarah's Key* (2011).

Given the near ubiquity of hospitality in cinematic representations of genocide, it is perhaps surprising that the discipline and practice of IR so often ignores it as an ethical possibility. For instance, the UK's miserly hostipitality to Syrian refugees mentioned in the Introduction reprises its earlier reaction to Kosovans in 1999 (Bulley, 2010). Such rejection of refugees is generally accompanied by calls for more international aid, often alongside military intervention. What links aid and military intervention is that, as 'ethical' practices, they tend to keep the other where they *belong*, as far away from 'us' as possible. Once 'we' can be sure 'they' won't appear at our door, we can feel both morally outraged at the evil acts of tyrants (substitute Assad for Milosevic, the Taliban, and so on), and content that we are doing all we can to help. Intervention and aid contain a problem elsewhere, limiting the need for hospitality. They make it safe again to think and act ethically (Orford, 2003). IR and international ethics have followed this tendency in theorising state responsibilities. For instance, both Nick Wheeler (1999) and Michael Walzer (2006) agonise over the definition of events that 'shock the moral conscience of mankind' and therefore warrant military intervention. But Wheeler does not mention hospitality as a way to 'save strangers'; Walzer sees hospitality as merely an imperfect duty of 'mutual aid' (1983: 21). Here, it is the statist ontopology of IR and ethics, the way they link belonging, identity and responsibility with the territorial state which severely limits their ethical imagination (Campbell, 1998; Bulley, 2006). They cannot *think* hospitality.

In contrast, cinema has had no problem in imag(in)ing hospitality as an ethical response to acts that shock our moral conscience, frequently highlighting Western states' dereliction of duty in this regard. Some suggest that film can speak to the issue of genocide in a way that other media cannot (Wilson and Crowder-Taraborrelli, 2012: viii). I am suspicious of such claims, but cinema has become a central way in which many confront genocide from the safety of their homes – *Hotel Rwanda* in particular has 'created a Western hegemonic discourse of the genocide' in Rwanda (Dokotum, 2013: 134). The central claim of this chapter is that, while being frequently ignored in the study and practice of IR, hospitality continues to be a crucial way in which popular culture envisages ethics when confronted by crisis. However, as I look at each film it becomes clear that, even in these idealisations of simply ethical acts, power emerges to disrupt these practices of hospitality as they construct the different spaces of homes, hotels and homelands. Ultimately, these romanticised pictures present hospitality as dependent upon the denial and erasure of difference which it is seeking to protect against. In a minimalist form, this hospitality mirrors the 'cleansing' it opposes.

## SOVEREIGN SILENCING: *HOTEL RWANDA*

*Hotel Rwanda* is based on events in Rwanda in 1994, when around 800,000 Tutsis and moderate Hutus were slaughtered in a genocide orchestrated by extremist Hutus and carried out by the *interahamwe* militias and Rwandan army. While there is a long history of massacres by both Tutsis and Hutus in the Great Lakes region, the film has been widely criticised for offering no historical or social context to the genocide (Adhikari, 2008; Glover, 2010; Khor, 2011; Dokotum, 2013). Instead, it gives a reductive account concentrating on the 'true story'[1] of one man, local hotel manager Paul Rusesabagina, and his heroic act of hospitality which saved 1,268 Tutsis and moderate Hutus by sheltering them in the Hôtel des Milles Collines. This distillation of the genocide to the acts of one heroic host was a very deliberate ploy to entice Western viewers and reach as wide an audience as possible. While Keir Pearson's original script contained an ensemble of characters, the director and co-writer Terry George asked for it to be 'pared down', focusing on Paul and his wife, Tatiana, so that the 'horrific element' could be reduced and it would also work as a 'love story'.[2]

As Debjani Ganguly (2007: 60) notes, 'In *Hotel Rwanda*, 'hospitality' is marked in literal terms by making the hotel, Des Milles Collines, the main site of action and refuge, and the hotel manager its protagonist'. Using hospitality as an ethical and structural organising device allows George to both retain a claim to authenticity while limiting audience exposure to the horror and bloodshed outside the hotel compound. A distancing is also achieved in the style of the film which, shot largely in Johannesburg (rather than Kigali), is harshly lit and given a Hollywood gloss in its production which rarely emphasises the darkness of tone seen in most genocide films (Foundas, 2004). This gloss is broken only once, where the starkness of the genocide is revealed as Paul finds that his van is driving over a road paved with dead bodies. Yet, the picture is softened at the same time: the scene occurs at night, in the middle of a fog, shielding the audience from its full horror. Nonetheless, the outside and its implied anarchy and violence constitutes the inside of the hotel as a space of relative safety and protection.

The focus on Paul and his hospitable heroism draws the inevitable parallel with *Schindler's List* (Adhikari, 2008: 177; Glover, 2010: 96; Gudehus et al., 2010: 352; Uraizee, 2010: 16), a correspondence which George directly invokes in several scenes (Hron, 2012: 140). Nearly everything in the film is shown from Paul's perspective, from the opening scenes where we are introduced to him as a shrewd, intelligent businessman, to the final scenes where he is victoriously reunited with his brother-in-law's children in a refugee camp. We see his journey from a commercial host, concerned to make money for his employer and protect his family, into a genuinely ethical host and saviour of over a thousand people. Paul is presented early on as a very recognisable figure that Western audiences automatically warm to, someone with fine taste in food, drink and suits, who knows how to deal with people and lives in a large house in a wealthy area of Kigali with his beautiful wife and children.[3] When asked to join the Hutu cause, Paul replies that he has no time for politics – he is a businessman, who deals with unsavoury Hutus because it is 'just business', making use of 'Hutu power' slogans to save his Tutsi bus driver in the opening scene. He is a pragmatist, negotiating his identity and that of his family in order to navigate the complexities of a country that is beginning to unravel along 'ethnic' lines.[4]

Initially, Paul's shrewdness leaves him unwilling to help those in need. When his neighbour is denounced as a collaborator, beaten and taken away by the army, Paul merely watches from the gate of his compound. His wife asks him to help, but he refuses, limiting his own sense of moral responsibility: 'all day long I work to please this officer, that diplomat, some tourist, to store up favours so if there is a time when we need help I have powerful people I can call upon.'

While the man was a good neighbour, 'he is not family. Family is all that matters.' This firm delineation of obligations gradually erodes as violence descends and Paul returns home to find a group of Tutsi neighbours already hiding in his home. They have become his guests without invitation; as Tatiana explains, he is the only Hutu they trust. When soldiers arrive at his home, itself undefendable from the descending anarchy, he makes use of his contacts in the army (those he has bribed) to negotiate safe passage for him and his family (now much enlarged by neighbours) to the only place that is, the hotel. He subsequently pays 100,000 francs of the hotel's money to the soldiers in order to save their lives.

Arriving at the Milles Collines, Paul is handed the keys by his white boss who is fleeing – symbolically and literally establishing him as the host. He is put in complete charge of the hotel, exercising this power by welcoming Tutsis as guests, establishing it as a space of protective hospitality. As host, he has to juggle a range of responsibilities and roles: facilitating the departure of his white guests; secretly hosting the non-paying Tutsis; bribing the army to protect the hotel; dealing with the hotel's owners in Belgium who are thinking of closing it; and managing the staff who challenge his authority (especially the Hutu porter who has taken up residence in the Presidential suite while threatening to expose the traitorous hospitality Paul is offering to the Tutsi 'cockroaches').

Meanwhile, more and more groups of displaced Tutsis begin to arrive as the UN officer tries to maintain order outside the gates. As it becomes clear that the UN will offer no salvation, Paul is made to realise that he has been betrayed by the European diplomats he trusted, cultivated and sought to emulate in a replication of past colonial relations – 'I have no history; I have no memory; I am a fool'. Nonetheless, when all seems lost and the Rwandan army orders all his guests to be killed, Paul is able to muster his contacts one last time through a call to the hotel's Belgian corporate owners, achieving a stay of execution. He then teaches his guests to do the same, using their international contacts to secure exit visas and safe passage. In his greatest act of self-sacrifice, he abandons his family who have acquired places in an ill-fated UN evacuation, staying behind to care for the guests that he, as host, cannot abandon to their fate. Finally, when all seems lost, Paul and the Milles Collines survivors are taken by the UN to a safe camp in rebel territory. This idealised ending ignores the squalidness, danger and violence of the camp, as well as Paul's subsequent return to Kigali (run by the rebel Rwandan Patriotic Front and now President Paul Kagame) before being forced to flee and seek asylum in Belgium (Rusesabagina, 2007: 210–243). Tying things up in this neater fashion helps assure Western audiences that, even in situations of unimaginable horror, human goodness, hope and survival is possible (at the expense of concentrating on the 800,000 who did not survive).

Paul's journey from 'selfish' businessman to 'altruistic hero' (Adhikari, 2008: 192–193), from commercial to ethical host, appears both complete and completely benign by the time the film's credits role. This is apparent from the film's marketing tagline: 'When the world closed its eyes, he opened his arms.' Yet, even such an obviously 'good' act of hospitality demands that power be exercised and any power relations muddy the waters of ethical purity. In many ways, Paul appears to enact the characteristic sovereign mastery that Derrida claims hospitality demands, as outlined in the Introduction. He emerges as the 'master' of the Milles Collines, 'assured of his sovereignty over the space and goods he offers or opens to the other as stranger' (Derrida, 2000: 14). The fact that this is not his 'home', but a commercial space, a 'hotel', is irrelevant because the film emphasises that this is Paul's way-of-being, his ethos – he is first and foremost a businessman, a manager. While he cannot control the boundaries of his house, the hotel is different. And as his ethos and the purpose of his hospitality changes, so does the space of the hotel.

We thus see Paul established as the Milles Collines' sovereign master as he arrives and is handed the keys that control the internal and external borders of the space. Paul requests a faxed confirmation of his mastery from the corporate owners of the hotel, which he then uses to build his authority with the Milles Collines staff. He alone is making decisions (in an early scene he tells Tatiana to 'leave these things to my good judgement'), controlling the hotel's boundaries by requesting protective troops to be stationed at its gates from the UN officer and the General he has cultivated. He must negotiate the presence of the enemy within, represented by the Hutu porter who challenges his authority, usurping and seizing parts of the hotel and spying for the genocidaires. It is Paul who decides to welcome in more 'refugees' as they appear from other spaces (UN-controlled camps, churches and schools) which cannot provide protection. He supplies the food and drinks for his guests by continuing to trade with the *interahamwe* and authorises the use of the swimming pool water for drinking and cooking.

By making Paul the heroic focus, George further underlines his ability to exercise a sovereign power. George has been criticised for the 'superimposition of a Western concept of heroism in representing an African experience' (Dokotum, 2013: 147). Paul is the 'exceptional African', a commercial manager able to rise above the typical African backwardness, violence and ethnic hatreds to reason like a Westerner (Glover, 2010: 104–105). He is thus the great, semi-Western saviour and protector. The director's commentary in the DVD extras involves George interviewing the real Rusesabagina on the on-screen events; speaking over one of the many scenes in which he saves the Tutsi guests from the Rwandan army, Rusesabagina sums up his role that is endorsed throughout the film: 'The only option they [the genocidaires] had, in order to kill the refugees, they first of all had to kill me. Because I was the only person who was protecting the refugees. So before they kill those refugees, the only solution was to kill their protector, who was myself.'[5] The sovereign host is, first and foremost, the protector of his guest and Paul's ingenuity in this respect is emphasised as he uses his unique abilities to negotiate and communicate with the outside. A by-product of the heroic sovereign mastery is that the guests around Paul are reduced to a silent mass. The 1,268 guests are rendered bit-part characters and extras with snatched lines of dialogue, most often forming a backdrop to the hero: appearing as huddled groups without agency, speech or choice. They stand behind Paul looking forlorn when the UN evacuation convoy leaves the hotel compound; they crouch on the ground awaiting his decision when he is ordered to shoot them at the hotel gates; they sleep in the hotel's corridors as Paul bustles through, all action, with the weight of the world on his shoulders. In the only real moment of guest-agency, we see a group of girls performing a Rwandan dance with smiling faces as Paul merely looks on. Here we see a hint of normal life and communal existence among the guests that does not purely revolve around Paul, though it still falls under his sovereign gaze and is enabled by his mastery of the hotel's boundaries.

It is Paul's moral journey as host above the fray of African backwardness that produces the Hotel des Milles Collines as a particular type of space. When we first see it the hotel is very clearly a commercial space, humming with Western guests, diplomats and army personnel going about their business and being charmed by Paul's offer of drinks and food. However, as the violence descends, the trajectories that produce the space change markedly, with a queue of white Westerners leaving and masses of Tutsis arriving as a result of Paul's sovereign decision to offer salvationary hospitality. The space becomes more crowded and messy, though in keeping with its glossing of the genocide, *Hotel Rwanda* 'conspicuously underplays' the misery and squalor of conditions towards the end of the 76 days, 'often tak[ing] on the aspect of a somewhat crowded holiday camp' (Adhikari, 2008: 186–187). Nonetheless, we do see communal eating, crowded rooms, the swimming pool water being used for cooking and cleaning, corridors full of sleeping people.

Moreover, this space is one that constrains movement. 'Guests' cannot leave because of the multiple dangers outside (shown in flashes of colour through the bamboo perimeter fence, short television reports, brief images from the windows of Paul's passing van, army and *interahamwe* visits to the hotel compound) that help to constitute the hotel as a space of safety and protection. Through Paul's changing practice of hospitality the hotel becomes a space for the maintenance of a bare form of life, vulnerable, exposed and without political voice or agency. The film thereby follows what Madeleine Hron (2012: 140) calls

> the generic narrative structure of the Holocaust texts: movement from the home to the "concentration camp," arrival in the camp, conditions in the camp, episodic killing growing in intensity, and eventual escape or liberation. Notably, most of the action in these films is confined to "concentration camp" like spaces – be it the Hotel des Milles Collines, schools, or churches.

Nonetheless, as the UN Officer arrives at the hotel early in the crisis with more 'guests', Paul firmly stresses that 'This is *not* a refugee camp'. Indeed, the hospitable space is neither a refugee nor a concentration camp, and part of Paul's ingenuity as host is the fact that he manages to produce a form of hybrid space. His heroic life-saving hospitality can only last as long as its commercial façade. In a speech to staff and guests alike he urges that 'Most importantly, this cannot be a refugee camp. The Interahamwe believe that the Milles Collines is a four star Sabena hotel. That is the only thing keeping us alive.' His ethical generosity is commercialised as a tactic of resistance against the genocidaires: he prints bills for all the guests that he knows they will never pay, continues to dress in suits and maintains staff uniforms and standards, prints fake guest-lists to fool the killers and ensures hotel-like appearances are kept up. The space is neither fully a camp nor a hotel, but something negotiated through this particular practice of hospitality.

Of course, as *Hotel Rwanda*'s stylistic choices mean that the hospitable space is produced solely through the decisions and ingenuity of the sovereign host, little resistance can be offered by the other guests. Their resistance is catered for by their sovereign host. In a scene that helps produce the criticism that George softens the genocide (Adhikari, 2008: 186), turning mass slaughter into a love story, Paul manages to find a small part of the hotel (the roof) in which to have a tender meal with his wife. However, this is a form of resistance reserved entirely for the sovereign host; other guests eat their meals communally. And Paul uses the opportunity to suggest to Tatiana that, should the *interahamwe* make it into the hotel, she should use the roof in a rather different form of resistance: killing herself and their children quickly. Paul's exercising of sovereign mastery to offer life-saving benign hospitality thus works to produce a hybrid space that protects *and silences* his guests, reducing them to docile bodies without agency whose resistance is performed for them.

## PASTORAL FEMINISING: *WELCOME TO SARAJEVO*

*Welcome to Sarajevo* is set during the Bosnian conflict in the 1990s where around 100,000 people were killed and over 2 million displaced. More specifically, its context is the siege of Sarajevo which lasted for three and a half years from 1992 to 1995. Sarajevo was crucial to the Serb war effort and became the centre of media attention, humanitarian relief operations and the Serbian artillery, forming the longest and most internationalised siege of a city in modern history (Andreas, 2008: vii–6). Like the other films examined, *Welcome to Sarajevo* is based in or around a true story, in this case that of English ITN journalist Michael Nicholson and his bid to smuggle a young girl from an orphanage in Sarajevo back to the UK. The film is centred on two distinctive practices of hospitality: the dangerous welcome experienced by Western journalists in the besieged Sarajevo; and the benevolent and caring hospitality practiced by Henderson (Nicholson's cinematic alter-ego) towards Emira

(Natasha in real life). Meanwhile, there is a constant background reference to the failure of Western Europe's hospitality. Henderson's reports from Sarajevo (echoing Nicholson's own) condemn the inability of Europe's governments to evacuate Sarajevo's civilians and children. This is set inside the wider failure of the UN to forcibly intervene to stop the ethnic cleansing.

As the first major film about the Bosnian conflict (and probably still the best known to Western audiences), *Welcome to Sarajevo* was shot on location in Sarajevo, Croatia and Macedonia only a few months after peace had been declared. The narrative is told almost entirely from the perspective of the European saviour-host, Henderson, and his British and American journalist colleagues, themselves 'guests' in Sarajevo.[6] The only Bosnian character given even minor development is the ITN crew's Bosnian Muslim driver, Risto. While receiving a lukewarm critical and commercial response, the film has been strongly criticised for offering a highly reductive account of the conflict (Iordanova in Bennett, 2014: 44). Though paying lip-service to the complexity of the war, it essentially portrays two ethnically homogeneous groups: Bosnian Serbs (wholly bad) and Bosnian Muslims (largely good if inert). This simplification is best represented by the film's changing of the original subject of Henderson's hospitality from the Bosnian Croat Natasha, to the Bosnian Muslim Emira.

The film begins with a jaded set of journalists, we assume all veterans of previous wars, holed up in the Sarajevo Holiday Inn. Henderson is jolted from his impersonal coverage of the conflict when he reports a story about an orphanage housing homeless Bosnian Muslim children. His project becomes to 'get those children out of here' by putting them on UK television every night, letting them tell their own story in order to move Western publics into action. The failure of state-based hospitality is underlined by news footage of UK Prime Minister John Major arguing that it is better to care for children near their family and home than evacuate them to unfamiliar surroundings. Counterposed to such a failure of conventional international hospitality, in a moment of weakness Henderson inadvertently promises the nine-year-old Emira that he will help her get out of Sarajevo to England. Whereas Nicholson in fact went to the orphanage with the express intention of evacuating a child, Henderson's impulsive response makes his hospitality appear entirely righteous: 'what was a premeditated act in "real life" appears on film to be a spontaneous, more ethically motivated response to a call emanating from the other herself' (Molloy, 2000: 82).

As a private citizen, Henderson lacks the facilities to enact the hospitality he has promised. However, when an American aid worker, Nina, arrives with a bus to evacuate children, Henderson secures places on the bus for some of the orphanage's babies on the condition that they have relatives in Europe that they can claim to be visiting. Their safety is thus guaranteed by becoming temporary guests, using the aid machinery to evade the hostility of European states. However, Nina cannot take Emira as she 'doesn't have somewhere to go'; she lacks a host. Emira recalls Henderson's promise and he is moved once more to respond, asking Nina if he could act as host: 'if she were visiting me'. He must respond by fulfilling his promise of hospitality. Agreement is reached between Nina and Henderson (without Emira's input) that this will be a temporary hospitality, 'just for a short time; till the war is over'. To be truly ethical in this case, the hospitality must be conditional and time-limited.

However, the spontaneity of Henderson's hospitality does not mean that it is free from power relations. Unlike Paul in *Hotel Rwanda*, Henderson cannot exercise sovereign mastery as he is outside his own territory, a long way from the home he seeks to welcome the other into. Instead, he employs a form of what Foucault calls 'pastoral power', setting up a different relation between the host (Henderson) and the guest (Emira), whilst producing the sanctuary of 'home' as a very different space to that of the Hôtel des Milles Collines. According to Foucault, pastoral power is a forerunner of modern forms of governmentality. He sees it emerging from the pre-Christian Mediterranean East, but 'above all' through the Hebrews, and subsequently in the Christian East with the development of the pastorate (Foucault, 2007: 123). It is essentially a religious form of power that God exercises over his people, either directly or, more commonly, through a representative. The king or ruler becomes the shepherd of the people, leading them like a shepherd, or a pastor, leads a flock (Foucault, 2007: 124).

There are three key elements to early forms of pastoral power, and these are particularly relevant to the hospitality Henderson exercises. First, while sovereign mastery is exercised over a territory, or a home in the case of hospitality, pastoral power is exerted over a flock 'in its movement from one place to another. The shepherd's power is essentially exercised over a multiplicity in movement' (Foucault, 2007: 125). Second, 'pastoral power is fundamentally a beneficent power ... entirely defined by its beneficence; its only *raison d'être* is doing good', as its sole objective is the salvation of the flock. It is thus a selfless power of *care* (ibid.: 126–128). Finally, pastoral power is individualising – the shepherd directs and has responsibility for the whole flock, but he can only do this 'insofar as not a single sheep escapes him' (ibid.: 128). This produces what Foucault calls the paradox of the shepherd': on the one hand, the pastor must both 'keep his eye on all and on each, *omnes et singulatim*'; on the other, there is the problem of sacrifice – 'the sacrifice of himself for the whole of his flock, and the sacrifice of the whole of his flock for each of the sheep' (ibid.: 128). Thus, he must 'care for each and every member of the flock singly' (Golder, 2007: 165).

Each of these elements of pastoral power can be seen in the ethics and power exercised in the hospitable relation between Henderson and Emira. While not a religious story, Henderson plays the role of God's representative in the film's narrative. After all, the unwilling but capable saviour of Sarajevo is the UN and Western governments. Henderson and the other journalists are representatives of this jaded, compassion-fatigued West, with the call of the other jolting him out of his ethical malaise and prompting the act of hospitality. Thus Henderson becomes a relatively uncomplicated Hollywood hero; like the second stipulation above, his is a power of care, completely benign, beneficent and without ulterior motive; his act is one of pure hospitality aimed at providing safety and salvation. Equally, Henderson's hospitality is all about mobility; it requires power to be exercised over a 'multiplicity in movement'. A key dramatic sequence in the film's narrative arc comes with the evacuees' tense bus journey, shepherded by Henderson and Nina across Bosnia to Croatia and onwards to safe European homes. This flock must be guided past road blocks and UN bureaucracy; they must be fed, sheltered and cared for. Meanwhile, the reference point for Henderson and Emira is always the sheepfold at the end, Henderson's home. He uses the journey to educate Emira, showing her photographs of his (and later her) family, teaching her some English words and the basic geography of her new home, the concentric circles of belonging: 'in London', 'in England', 'in the UK'. The journey is crucial in producing, and preparing Emira for, the home that the audience has yet to see but will be created through this act of hospitality.

A crisis in this pastoral journey occurs when a Bosnian Serb road block allows a 'Chetnik' officer access to the list of children and he proceeds to remove those with Serbian names, including a baby that Emira has been looking after since we were introduced to her in the orphanage. Henderson and Nina must choose how to exercise their limited power in the face of violence and danger – do they sacrifice themselves and potentially their flock by fighting their case; or do they sacrifice these few for the flock's better protection? The paradox of the shepherd allows no satisfying ethical response, but Henderson's commitment is individualising and he helps hold Emira back, releasing her hands from round the baby and sacrificing it to the Chetniks. The appalling moral paradox remains in place, though the film leaves us in little doubt that Henderson exercised his power responsibly as the narrative quickly moves on. In exercising such pastoral power, however, Henderson places his own desires and calculations above those of Emira who is not offered a choice as to how the individualising nature of Henderson's salvationary hospitality is conducted.

The pastoral nature of this hospitality is also central to the production of Henderson's home. Shortly after the road block, the film shifts to grainy hand-held camera footage of Emira's first year in the UK, with family and friends, a beautiful home, fireworks, childish pleasure and Emira's own birthday party. The images display an idyllic vision of safety, with green lawns and trees awash with sunlight. However, this picture of home as 'heaven' is necessary because of the contrast with Sarajevo as 'hell'. While the producers made key changes to the 'true story', they worked hard to achieve authenticity and veracity in representing Sarajevo. Winterbottom makes extensive use of the war's immediacy: using the real debris of war as a backdrop, intercutting documentary and news footage with the film's own glossy production, and local Sarajevan eye-witnesses aiding in the reconstruction of key scenes. Thus, while the religious aspect of pastoral power might not be central to the hospitality offered, it is to the visual creation of the home. Matt Roth's (1998) review notes that:

> The second half of the movie, far from welcoming you into the besieged city, centers on a Dantean flight from it. A contrast is established between infernal Bosnia and paradisiacal England: Henderson's London is so Edenic it seems to consist entirely of sunny gardens and quiet plush interiors. The Hendersons are constantly shown in or around fluffy, inviting beds, a visual contrast to the dirty mattresses barricading the windows of Sarajevo apartments.

The security and comfort of the sheepfold home is generated through a stark contrast with the hell of Sarajevo and the trials and dangers of the pastoral journey to safety. In reality, this home itself was besieged shortly after their arrival as tabloid journalists heard of Nicholson's incredible (and illegal) hospitality (Nicholson, 1994). Such non-dichotomising details do not make it into the film, underlining the righteousness of Henderson's actions.

Something less benign becomes apparent in the highly gendered production of the spaces of 'home' and 'away' in *Welcome to Sarajevo*. The traditional masculinisation of war and conflict has been well demonstrated by feminist approaches to IR (Sylvester, 1994; Kinsella, 2005; Sjoberg, 2014). This is reproduced in the film as those doing the fighting are all men; those that Henderson seeks to save, both through his reporting and pastoral hospitality, are the classic feminised victims – babies and children. Similarly, while Henderson is the generous host, he is a very masculine host: searching for violence and conflict before shepherding the feminised other back to safety. Meanwhile, his wife, who will presumably do most of the *work* of hospitality, feeding, caring for and ultimately raising Emira while Henderson is searching for new conflicts, silently accepts being the fulfiller of her husband's generosity that was offered without consultation. The hostess, as in most hospitality narratives, becomes merely an extension of her husband's subjectivity and beneficence (McNulty, 2007: xxxvii). The first time Henderson informs his wife of his hospitality is a late-night phone call from Croatia, the day before they reach the UK.[7] Mercedes Maroto Camino (2005: 123) argues that this places the film within the cowboy Western genre, where heroic men are counterposed to two types of women, either '"the angel in the house", away from the fray' or 'whores'. Both types 'provide a background for the more important and interesting lives of men'. This gendering of the safe space created through hospitality and the work done there is reproduced in *Hotel Rwanda*. When Paul shows a female member of his staff a room full of sleeping Tutsi children he has saved, his only words are 'Bathe them; feed them; put them to bed'. Such caring duties are not the work of the (male) host.

The contrast between the spaces of 'home' and 'away' are further underlined in the final major crisis point of the film, and this is portrayed specifically through a contrast of mothering care and its production of the home. Literally interrupting the idyll of Henderson's hospitality, a phone call from his producer reveals that, though found in an orphanage, Emira has a mother who wants her back in Bosnia. Emira does not wish to return and Henderson must, like the good 'watchful' shepherd, return to Bosnia to find the mother and persuade her to give Emira up to his family's adoption. He meets Emira's uncle who tells him that Emira's mother was forced to give her up: 'She's lost so much. We've all lost in this war. If something could be returned that's ours, we'd like to have it back.' This completes the chain that connects Saraejvo to a violent masculine hell and its failure of family and motherly love, and in turn London to its idyllic feminised welcome and motherly devotion.

Risto drives Henderson to a heavily bombed area, searching for Emira's mother – 'Is this where they want her to live?' asks Risto – where they will be mugged and robbed, further emphasising the violence of Sarajevo. Risto makes Henderson promise not to bring Emira back. The promise is simply, unproblematically given, between two men; Emira's mother and her wishes are irrelevant. Risto will die the next day, underlining the straightforward beneficence of Henderson's decision. He eventually finds the mother with the help of a local gangster, and the following exchange ensues:

| | |
|---|---|
| *Emira's mother*: | Mr Henderson, I'm her mother. And I'm alone. I know that I was a bad mother. That I gave her up, but I love her. Do you understand why I want her back?' |
| *Henderson*: | Yes, I understand ... But umm ... Emira hardly knows you. You've seen her twice in eight years. And to be perfectly honest – |
| *Translator*: | I don't translate this. You must listen. |
| *Emira's mother*: | I am giving you Emira to be your daughter. But because I was a bad mother I don't have any memories of her. I want to see her. To hear her voice. |

Henderson shows her a home video of Emira's new life and calls Emira to speak to her mother. Emira is initially confused, forgetting that she has to speak in 'Bosnian' and preferring English. Then, speaking in her mother tongue to her mother she confirms her British belonging, her refusal of a guest identity and her becoming-host: 'I'm happy here. This is my home now. Goodbye.' Her mother immediately signs her adoption papers and we are directed to feel a 'grim satisfaction' (Roth, 1998) and little sympathy for the mother's failure of femininity and care. Similarly, perhaps we blame the ethnic cleansing on the failure of Bosnia and Sarajevo to produce proper homes that would not require Western intervention and pastoral hospitality.

Emira's subjectivity and identity have thus changed, even before the adoption, from that of 'guest' to that of belonging – London 'is my home now', not Sarajevo. The hospitable 'saving' of Emira is thus spatial and affective, but it is also *transformative*. This is given visual representation through the change in Emira's appearance from what Molloy (2000: 83) calls a 'tough-talking, cigarette-smoking, androgynous-looking child with short cropped hair' in dirty jeans and a T-shirt; in her new home she is a happy girl in a dress and sporting long hair. Emira is thus domesticated to a feminised form of belonging that is necessary to her new status as British, no longer a guest. Winterbottom stresses that in casting Emira they deliberately sought someone not too cute,[8] and this aids in her transformation; 'we' (the Western audience) warm to her by the end as definitively 'one of us'.

The totalising nature of the dichotomies that structure *Welcome to Sarejvo* and the apparently unquestionable rightness of Henderson's hospitality means that the film offers little by way of moral complexity. There appears no room for, or evidence of, resistance to its narrative. We do not see Emira struggle with her transformation, the traumas of moving from hell to heaven, from guest to host, from non-belonging to belonging. We do not see her grow up and rebel against her new identity and status.[9] And we certainly never hear from Emira herself. No reflection is offered on the ethics of the pastoral power exerted by Henderson's actions, or the violently gendered home it produces. However, we do see the tidiness of the narrative undo and resist itself through a final device used by all the films examined in this chapter. As it ends, but before the credits roll, a stark message appears on the screen in white against a black background enumerating how many people lost their homes in the conflict, how many died or went missing, how many children were killed. Finally: 'Emira still lives in England.' This last sentence inadvertently demonstrates the implausibility and impossibility of the ease with which Emira moves from guest to host. After all, why does this need stating? She was adopted and became one of Henderson's children. She became English; so why is her continued residence worthy of remark? This 'still' hints that the British director and producers continue to see her as a guest, in perpetuity, no matter how much her otherness has been transformed and effaced.

## GOVERNMENTALISED AMNESIA: *ARARAT*

In contrast to both the latter two films, *Ararat* explicitly breaks with the immediacy of violence, set, as it is, nearly a century after the Turkish genocide of Armenians in 1915. Just as *Welcome to Sarajevo* effaces Emira's change from guest to (almost) host, *Hotel Rwanda* tells us nothing about Paul's subsequent journey as a refugee, nor how his children coped with the trauma of their experiences. In *Ararat* we see how such hospitality can breed struggles of memory, adjustment and trauma across several generations that *Hotel Rwanda* and *Welcome to Sarajevo* both sacrifice to a tidy, linear narrative. A Franco-Canadian production directed by Atom Egoyan, *Ararat* also makes no claims to be based on a 'true story'. Rather, it plays on the ability to present 'truth' through cultural and artistic forms. The film's broken narrative structure circles around the siege of Van and the subsequent massacre of Armenians by Turkish Ottoman forces between April and May 1915. The representation of genocide, while horrific, is always distanced and mediated by being shown as a film within a film, with artistic embellishments constantly discussed, depicting the edges of the set and often including the film crew within the frame.

In its deviating, non-linear structure, *Ararat* performatively demonstrates the fracturing and impossibility of clear representation and definitive memory. This makes it difficult to summarise as a coherent plot, moving back and forth temporally and spatially between several different intertwining stories. One story follows the film within a film (also called *Ararat*) about the Siege of Van. This is largely what we would expect of genocide cinema: based on the eye-witness account of the American missionary (Clarence Ussher) in charge of the US mission compound inside the Armenian quarter of Van, where many Armenians sought sanctuary from the genocide. A second story follows the life of the Armenian-American artist Arshile Gorky, told using interspersed flashbacks, shot in the same stylised fashion as the film within a film. These two stories collide through artifice as the director and writer of the film within a film decide to put Gorky into this story when they find out that Gorky was a child and possibly present (though certainly not as the writer includes him) during the siege. They also intrude on each other as ways of remembering genocide, with the Gorky expert-consultant interrupting a key scene from the film within a film as Ussher, the great white hero, attempts to save a dying baby in the mission compound. Art is used to both construct and deconstruct the possibility of a neutral space to remember, account for the past and welcome the other.

However, these stories of the genocide are not central. Rather, they form a background to the main two plots which circle around Raffi and David, whose meeting a year after the film within a film has been shot structures *Ararat*. Raffi is an Armenian Canadian production assistant on the film, struggling with his family's past and the contending identities and loyalties he owes to a liberal, multicultural Canadian homeland, and the Armenian homeland of his forebears. His father was a freedom fighter/terrorist, killed trying to assassinate a Turkish diplomat. David is an elderly Canadian customs official working at a Toronto airport. He is struggling to accept his son Philip's homosexual relationship with a Turkish Canadian actor, Ali. In one of the many intertwinings, Ali will play the evil, genocidal Turkish officer, Djevdet Bey, who presides over the siege at Van.

This convoluted web of narratives helped contribute to the film's poor critical reception, with reviewers condemning its 'overly contrived' structure (Holden, 2002), finding it 'needlessly confusing … too heavily layered, too needlessly difficult, too opaque' (Ebert, 2002), and a 'bafflingly flawed project' (Bradshaw, 2003). While *Ararat* provoked predictable outrage in Turkey, many Armenian-Canadians were also unhappy that the film's complexity diluted the impact of a black-and-white issue (Romney, 2003: 184). Egoyan answers this by arguing that *Ararat* is not really a film about the Armenian genocide; rather, it is a 'film about living with the effects of the denial of that event into the present' (in Romney, 2003: 173). It is therefore more a film about subsequent generations negotiating their coexistence and memory of genocide in a country that is not Armenian, Turkish or Ottoman. Thereby, 'the strained relationships in *Ararat* silence the diasporic communities as they struggle to articulate their own testimony and their refusal of Others' testimony' (Frieze, 2008: 244).

In following these various characters and their coming into relation by circling around the artistic creation of a version of the past, we see something that no other genocide film shows us: the negotiation of ethics, ways of being in relation to each other, as time and distance blurr and (re)animate the memory of it. A key process by which these relations occur is through unsuccessful and partially successful acts of hospitality that help draw the boundaries and borders of the Canadian and Armenian homelands. Where the genocidal hospitality in *Hotel Rwanda* and *Welcome to Sarajevo* appears comparatively simple and straightforward, producing feminised homes and protective hotels, the various practices of hospitality represented in *Ararat* construct complex and contested spaces. Here, the space of hospitality is constantly being interrupted and redrawn by the incursions of memory, artifice, difference and operations of governmentality. I will concentrate on two spaces in particular here: the entrance to Ali's home and the Canadian border at the Toronto airport.

The first scene occurs after Ali, who plays the evil Turkish general and is half-Turkish himself, has tried to dispute the veracity of the genocide with the film's director, Edward Saroyan. Raffi is dispatched to drive Ali home and Saroyan tells Raffi to give him a bottle of champagne. Raffi, unable to accept Saroyan's equanimity, continues the debate outside Ali's flat, but Ali flatly rejects the importance of remembering the past:

> Look, I was born here. So were you, right? This is a new country. So let's just drop the fucking history and get on with it. No one's gonna wreck your home; no one's gonna destroy your family. So, let's go inside and uncork this thing [a bottle of champagne] and celebrate.

In a scene that plays out the possibilities and limits of a neutral Canadian multiculturalist ethos that both encourages and contains difference, this offer of hospitality is refused by Raffi. He recalls to Ali the apocryphal tale that Hitler convinced his commanders of history's forgetting their actions by saying 'Who remembers the genocide of the Armenians?'. Ali replies, 'And nobody did. Nobody does', before retreating alone into his home. Ali has become a sovereign host, primarily a Canadian rather than a Turkish-Canadian, setting the terms upon which his welcome can be offered and accepted. His own place in the 'new country' (it is 'new' to neither Raffi nor Ali) is secured precisely through his forgetting. To be welcome in Ali's home and the Canadian homeland, dangerous forms of difference (such as memories of genocide) must be disavowed, covered over by the celebration of sameness. In a similar way to Emira, who became a long-haired, dress-wearing, British girl who forgot her mother-tongue, Raffi must perform his homogeneity, or tamed difference, to be welcome. The liberal, open and tolerant hospitality offered by Ali and Canada is itself minimally genocidal, as it seeks to erase forms of otherness deemed threatening to the 'home'. But Raffi refuses this offer of hospitality and its conditionality (that he forget), having been warned by his girlfriend, mother and Saroyan of the importance of remembering. He remains a guest in the only homeland (Canada) he has ever truly known precisely because he refuses to forget.

Raffi's 'guestness' is reinforced and ultimately reaches some kind of resolution through his meeting with David in the central encounter of the film at the Canadian airport border, which we eventually realise gives the intertwining narratives their structure. This is the second space of hospitality I want to explore, as it builds upon the failed welcome at Ali's flat. The meeting takes place as Raffi reaches Toronto, returning from a trip to Turkey. Canada as the most important home/homeland in the film has already been emphasised by the first scene in *Ararat*, where Saroyan confronts David whilst entering the country. Saroyan's bag is checked by David and, finding a pomegranate, a symbol of Armenian culture, he is informed that it cannot be brought into Canada. Saroyan responds by cutting open the pomegranate and eating it, offering some to David while telling the story of how the fruit reminds him of his mother. The nominal guest (Saroyan), in the act of having conditions placed on his welcome, reverses the relationship with the host (David) by trying to feed him and fulfil the role of host. He thereby also draws attention to the internalisation of memory and its uncontainable, uncontrollable nature: the difference contained in memory can be smuggled across the border through internalisation (the eating of the pomegranate).

When David, as the white, male representative of the Canadian state, searches Raffi at the border he finds that Raffi has returned with a set of sealed cans of exposed film, which Raffi claims contain process footage for the film within a film. Due to the non-linear nature of the narrative, the audience initially has no reason to think this is untrue. David, the old pro who is about to retire, is suspicious. Instead of welcoming him back to his homeland, David questions Raffi throughout the night after physically checking his arms for drug use. Raffi is forced to give an account of the film, the genocide and its politics, his family and, ultimately, himself, his identity and history back to AD 451 (when Armenians repelled the Persians). While we, the audience, do not know what is in the cans (opening would expose them to the light and destroy the film), Raffi's story continually changes as David peels away layers of untruth. As the other narratives progress and Raffi's identity is pulled apart by his step-sister/girlfriend, his mother, Saroyan and Ali, *Ararat* keeps circling back to this structuring encounter with David in which his account of himself also gradually falls apart.

The power David exercises here appears to be that of the sovereign master, similar (though neither heroic nor benign, as our sympathies are with Raffi) to that of Paul in *Hotel Rwanda*. It is his job to determine who/what is welcomed and who/what excluded. This appears to be reaffirmed by David's ability to question and demand answers. Yet these encounters with Saroyan and Raffi ultimately demonstrate that the power David exercises as host is more nuanced than sovereign mastery, as is the homeland that hospitality produces. Rather than just seeking to police the border through inclusion and exclusion, David's tactics are part of a wider form of governmentality. This mode of power has been characterised by Colin Gordon (1991: 5) as the 'conduct of conduct', the various means by which actors' behaviour and actions can be controlled (Foucault, 1982: 791). Instead of classifying the welcome and the unwelcome, governmentalised hospitality is about the variety of ways in which actors seek to produce people and populations, while maintaining, managing, controlling and directing their behaviour (Butler, 2004: 52). Governmentality operates via a liberalising logic of security, the aim of which is to sustain and optimise life. Security here does not mean complete invulnerability, but rather aims to promote the circulation of 'good' subjects and objects, while restricting the mobility of the risky (Foucault, 2004: 246). What we see at the border in *Ararat* is David's part in processes of governmentality that filter out risky mobilities (the pomegranate; the film cans; perhaps Raffi) and ensure the continued mobility of the safe (Saroyan and, perhaps, Raffi). In doing so, such hospitality produces Canada as a liberal space, neutral with regard to everything but its own security, the optimisation of the multicultural Candian way of life.

The sticking point comes with Raffi himself; precisely how risky is he as a subject? After all, as David observes, he has 'no way of confirming a single word' of Raffi's story. To determine this, David exposes him, his history, family, identity and memory to a seemingly endless interrogation, even watching the personal footage Raffi recorded on camcorder for his mother of Armenian relics near Van. He constructs a variety of identities for Raffi, all of which are more or less threatening for the Canadian homeland: a liar; the 'son of a terrorist'; the stepbrother and boyfriend of a drug dealer, credit fraudster and vandal. Yet, David also offers understanding of Raffi's pilgrimage to Turkey (to 'find meaning' he has lost), and Raffi's account of the horrors of the genocide and the Armenians' plight gradually unsettle David's confidence in his own knowledge. This is partly due to David's struggles with controlling his own behaviour as a guest, specifically his homophobia. In an early scene, his son Philip tells David that being welcomed in Philip and Ali's home depends upon him controlling his illiberal views: 'Either you make an effort to change your attitude … or you're not welcome at our place anymore.'

Finally, Raffi reveals that a guide who had helped him gain footage of Mount Ararat asked him to carry the film cans into Canada. Raffi asks that the cans be opened in the dark to save the film reels he still somehow believes are inside. As the lights go out, David can no longer see Raffi. His identity has been totally erased. While opening them in darkness, David asks Raffi why he believes the cans contain film, to which Raffi replies: 'Because if I didn't I'd be a criminal.'

Exposure to the light will therefore ultimately resolve and confirm his identity and status of belonging one way or the other. As Raffi leaves the airport, having been allowed into the country by David, a cut back to the room reveals that the cans contained heroin. Later, Philip will ask David why he let Raffi go.

*David*: I trusted him.

*Philip*: But he was lying to you all night …

*David*: The more he told the closer he came to the truth, till he finally told it. I couldn't punish him for being honest.

*Philip*: But he was smuggling drugs.

*David*: He didn't think he was.

*Philip*: How do you know?

*David*: He didn't believe he could do something like that.

*Philip*: Dad, what came over you?

*David*: You did, Philip. I was thinking of you.

This romanticised conclusion reveals how both David and Raffi's conduct has been conducted, as host and as guest. Raffi has found a resolution, found meaning through his pilgrimage and re-entry. He is now ready to manage his own behaviour as a less risky liberal subject (otherwise David would have had him arrested), acceptable to the neutral Canadian homeland. Yet he has not relinquished his dangerous and irruptive memories of genocide. Rather, he has opted to carry them internally (like Saroyan's pomegranate), smuggling them across the border in a relatively unsubversive form of counter-conduct. David's actions meanwhile appear benign, liberal, generous and hospitable. He has become the ethical hero of this governmentalised hospitality, having gone on a journey from bigot to host comparable to Paul's in *Hotel Rwanda* (commercial to sovereign host) or Henderson's in *Welcome to Sarajevo* (jaded journalist to pastoral host). Yet his decision to welcome Raffi also tells us something about how power is producing Canada here. David did not follow the governmental 'rules', but used the discretion of his position to determine Raffi as non-risky despite the 'fact' that he *is* a criminal, a drug smuggler. While exercised in an enabling rather that constraining manner, this is what Judith Butler calls 'the resurgence of sovereignty within the field of governmentality' (2004: 56). David's decision is unilateral and refers to no law or legitimate authority. He knows Raffi, the potentially dangerous element, better than Raffi does; David knows from the start he is both a criminal and not a criminal whereas Raffi leaves unaware that he is, by law, a criminal.

As a representative of the Canadian state, David's decision works within *Ararat* to affirm the 'new country' as liberal, hospitable, tolerant and neutral to difference. Yet two elements work against this production. First, we do not know for sure what kind of a resolution Raffi has reached. Georgiana Banita (2012: 89–90) argues that in *Ararat,* Egoyan is suggesting an 'ethical approach to memory recuperation and reconstruction' best exemplified through Raffi's journey. But has Raffi merely internalised his memories, like Saroyan's pomegranate? Or has he relinquished them? Is his acceptance of David's decision effectively an acceptance of Ali's in a re-run of that failed hospitality? Has he agreed to, like Emira, sublimate, forget and erase the dangerous aspects of his difference in a true acceptance of genocidal hospitality? And second, though David's exercising of a petty sovereignty appears generous and open-handed, it is evidence of the 'ghostly and forceful resurgence of sovereignty' within processes of governmentality, identified by Butler (2004: 59) in the use of indefinite detention of terror suspects after 9/11. Far from benign, David's discretion is merely the flip-side to the border guard's ability to criminalise the legal, to '"deem" someone dangerous and constitute them effectively as such' (ibid.: 59), leading to detainment without legitimacy or legality. Suddenly this ethical and welcoming hospitality appears not so far from tyranny.

## CONCLUSION

In a world in which it is generally agreed that hospitality is an ethical good and yet states deny a welcome to those suffering the worst forms of genocidal and ethnic persecution, cinema's reimagination of ethical possibility can provide welcome relief. Here, hospitality is not only the right thing to do; its successful practice acts as an ethical corrective, turning violence into heroic entertainment. This has prompted a new genre of genocide cinema (Wilson and Crowder-Taraborrelli, 2012), where hospitality often takes centre stage. Yet, when we examine the power relations which enable this hospitality, the subjects it produces and the particular kinds of spaces it generates, we are able to see the contingencies and dangers in cinema's imagined ethics of post-sovereignty.

Even in their idealised form, then, practices of hospitality work to efface and 'cleanse' difference. The protective practices of sovereign hospitality we see portrayed in *Hotel Rwanda* work through a Westernised hero reducing those 'saved' to a huddled mass of silent, dependent guests. The hotel, constituted by the sovereign host as all that stands between guests and the violence outside, changes from a commercial, cosmopolitan space to a hybrid space of safety, fear and restricted mobility. Here, a commercial façade allows the continuation of protection at the expense of the action and resistance of the guests. Belonging is premised on the relinquishing of agency. Meanwhile, the hospitality afforded to Emira in *Welcome to Sarajevo* is peculiarly pastoral and patriarchal. She herself has little say beyond eliciting the response of the male host and his promise of welcome, subsequently being shepherded to the safety of a new home. This secure, loving, feminised family home in London is organised in opposition to the masculinised hell of the besieged Sarajevo. But long-term hospitality and belonging is premised on the renouncing of all signs of otherness: language, short hair, cigarette-smoking, masculine clothing and any sense of trauma. Hospitality has achieved what the ethnic cleansing of Sarajevo could not. *Ararat* is far more complex in this regard, using genocide as a background to 'staging the predicament faced by many Canadians of remembering their cultural roots while not being incapacitated by this memory' (Banita, 2012: 101). Ultimately, however, the construction of an open, tolerant, multicultural Canadian homeland is premised on forgetting or pacifying dangerous memories of violence, trauma and difference. Such difference is managed and controlled, taught to adapt itself to acceptable forms that are non-homophobic, non-criminal and non-terrorising. But ultimately, the conditions placed on this hospitality render the difference of the guest unthreatening, neither irruptive nor disruptive.

If power relations are evident even in the idealised forms of ethical hospitality displayed in cinema, our attention is therefore directed back to its more common, everyday forms in the concrete relations of international politics. While *Ararat* gives some indication of how the difference of the guest is managed and this management resisted, it also ultimately returns to the state as the space of hospitality, whether or not as a cosmopolitan homeland. In the next chapter, I turn to one of the increasingly common spaces produced to cope with the everyday displacement of people from their homes and homelands: the controversial and contested space of refugee camps.

## NOTES

1 The 'truth' of *Hotel Rwanda* is contested. The script relies almost entirely on the testimony and memories of the film's hero, Paul Rusesabagina, which is disputed by many, including some of the 'guests' he saved (King, 2010: 299). The complexity of the 'truth' of this story is increased because Rusesabagina's memoir (2007) is in large part based on the screenplay for the film which came first (Dokotum, 2013), unlike the more conventional relation we see in the conversion of *Natasha's Story* into *Welcome to Sarajevo*.
2 See interview in 'The making of *Hotel Rwanda*', part of DVD 'extras' on 2005 release.
3 It is intentionally presented as a 'very Western lifestyle, not unlike people in Europe or America' (George in DVD extras).
4 The constructed nature of this 'ethnicity' is demonstrated through the narrative technique of informing outsiders (Western journalists) that the difference is a colonial hangover from Belgian's division of 'types' of Rwandans in order to rule them more easily. This well-meant account is both simplified and factually incorrect (see Glover, 2010; Dokotum, 2013).

5 See 'Director's commentary' – DVD extras. Rusesabagina's real role in this respect is widely thought to have been massively overstated, both in *Hotel Rwanda* and his own self-aggrandizing (Ndahiro and Rutazibwa, 2008; Caplan, 2009; Dokotum, 2013).
6 Though purporting to be about Sarajevo, the film focuses entirely on Western journalists. As such, it echoes Nicholson's book on which the story is based, *Natasha's Story* (1994), in which Natasha plays a very minor role.
7 This scene appears to be a fairly accurate representation of the phone call Nicholson himself made to his wife. While she did accept his decision, Nicholson reflects at some length later in his book on the difficulties she experienced and the unequal division of labour his unilateral hospitality forced on his wife (see Nicholson, 1994).
8 See 'Interviews with cast and crew' in DVD extras.
9 In *Natasha's Story,* by contrast, Nicholson (1994) offers fascinatingly paternalistic accounts of Natasha's early struggles.

# 2

# Humanitarian Hospitality: Refugee Camps

Forced migration is part of the everyday fabric of international politics. Whether the cause is civil war, environmental degradation, political, economic or social turmoil, having to leave one's home and rely on the hospitality of others is devastatingly familiar. According to the UNHCR (2015: 2), an average of 42,500 people per day were forced to leave their homes and seek hospitality elsewhere in 2014.[1] By the end of that year the total number of displaced people worldwide had risen to a record 59.5 million, the highest since records began, an increase of over 14 million on 2012 (UNHCR, 2013). Their population is now somewhere between that of South Africa and the UK and includes 19.5 million refugees and 1.8 million asylum seekers (UNHCR, 2015: 2). However, despite the concentration of hospitality literature on the immigration, asylum and sanctuary policies of Western states and cities (Derrida, 2001; Carens, 2003; Benhabib, 2004; Darling, 2009, 2011, 2013; Baker, 2011; Bagelman, 2013; Squire and Darling, 2013), the vast majority of the displaced remain in the global south, or 'developing world', including almost all internally displaced persons (IDPs) and over 86 per cent of refugees (UNHCR, 2015: 2).

In this context, offering hospitality to the displaced in the form of (more or less) temporary camps has become a common fix, alongside the UNHCR's acceptable long-term solutions of repatriation, integration or resettlement in a third country (2007: 44). While most Syrian refugees have not sought the hospitality of camps, in Sub-Saharan Africa 64 per cent of refugees did so in 2013, a figure that is steadily rising (UNHCR, 2014b: 37). Such transitional settlements are quite literally spaces created to offer hospitality to those that need it most, giving food, shelter and (some) security. The welcome provided by camps is not merely palliative; it aims beyond mere survival, allowing recovery, providing dignity and sustaining 'goods' such as family and community

(Sphere Project, 2011: 244). 'Camps replicate an entire support system' (Corsellis and Vitale, 2005: 115). This makes refugee camps appear the most obvious of the spaces I explore in this book – unlike cities or postcolonial states, camps have no purpose other than hospitality and they seem a straightforward, if temporary and imperfect, ethical response to suffering.

The picture is actually far more complex. Camps are highly diverse and varied spaces, despite frequent attempts to render them in a generalising and reductive manner (e.g. Diken and Laustsen, 2005, 2006). Jennifer Hyndman (2000) argues that camps are in fact spaces of containment run by international agencies, closer to a prison than a community, a home or a 'humanitarian space' (Yamashita, 2004). The first part of this chapter will therefore tackle this issue, examining the elements that make camps difficult to view as spaces of hospitality: their temporary nature, their insecurity, how they restrict refugee agency and the lack of a clear 'host' whose home is opened to the other. I argue that when we view hospitality as a spatial-affective relation of ethics and power (as outlined in the Introduction), refugee camps emerge as spaces produced by the hospitality of an ensemble host made up of a diverse range of international and local actors.

The second section examines precisely how camps are produced and managed as effects of an assembled host employing domopolitical governmentalities to better guide and shape the displaced guest's welcome. To illustrate this, I turn to three interrelated and cross-referential guides and handbooks on the spatial planning of these spaces: the UNHCR *Handbook for Emergencies* (2007), the Sphere Project's *Humanitarian Charter and Minimum Standards in Humanitarian Response* (2011)[2] and Oxfam/shelterproject's *Transitional Settlement: Displaced Populations* (Corsellis and Vitale, 2005).[3] Together, these guides form something of a manifesto or template for the governmentalised production and management of hospitable humanitarian spaces. Such conduct does not, however, reduce refugees to a voiceless biopolitical mass of bare life. Rather, as the final section demonstrates, refugees in an array of different camps have engaged in creative counter-conducts that subvert the host's control and seize the camp space, repurposing and reforming it. The government of hospitality that produces camps also generates a range of unintended consequences and contestations of the camp as a space and an ethos. These are illustrated using existing research by anthropologists and political geographers in several, very different camps across Africa and the Middle East.

## ASSEMBLED HOSTS AND HUMANITARIAN GOVERNMENT

It is very difficult to speak of refugee camps in general terms. All are produced through practices of hospitality, but they vary enormously across time and space. According to the UN, there are at least four major types of 'space' conceived as camps: 'cross-border points'; 'transit centres'; 'refugee camps' or 'refugee settlements'; and camps for 'internally displaced persons' (Agier, 2011: 39). Michel Agier argues for a different analytical typology which includes: 'self-organized refuges' such as ghettos, informal camp grounds, squats and border crossing points; 'sorting centres', such as holding centres, waiting zones, way stations and transit centres; 'spaces of confinement', such as refugee camps and UNHCR rural settlements; and 'unprotected reserves' (ibid.: 39–59). The analytical utility of such categories is open to question, but the need to disaggregate from the totalising concept of 'the camp' is clear: refugee-founded, administered and governed camps, such as those of the Western Saharan Sahrawi in Algeria (Herz, 2013) are not identical to state-controlled camps, such as the 'container city' for Syrian refugees near Killis in Turkey (McClelland, 2014).

Merely noting this variability is one way of dealing with some of the most generalising claims about *the* refugee camp which often take inspiration from the work of Giorgio Agamben (1995, 1998). Far from spaces of hospitality, such accounts tend to portray them as spaces of sovereign power and exception that sustain an agency-less bare life (see Perera, 2002b; Edkins, 2003a; Edkins and Pin Fat, 2005; Hyndman and Mountz, 2007; Orford, 2007; Hanafi and Long, 2010). At the furthest extreme, camps are portrayed as 'abstract spaces' or 'non-places' that 'fail to integrate other places, meanings, traditions' (Diken and Laustsen, 2005: 86, 2006; Diken, 2004). While perhaps true of some European and Australian asylum detention centres that are the focus of many such studies, this account of camp spaces is difficult to defend. Agamben-inspired work has thus been criticised for its depoliticising and agency-effacing effects from a variety of perspectives (Huysmans, 2008; Owens, 2008; Papadopoulos et al., 2008; Walters, 2008; Andrijasevic, 2010; Rygiel, 2012, 2011). As Adam Ramadan (2013: 68) observes, this reading is simply unsustainable as a general interpretation of contemporary refugee camps. Nevertheless, there remain profound difficulties in describing camps as spaces which are produced through practices of ethics as hospitality. This section will therefore tackle this issue, arguing that they can be read as such even though they have neither a sovereign host nor an uncontested ethos, and often fail to serve their basic function of providing security.

## ASSEMBLED HOSTING

If, as I have argued, hospitality is a spatial-affective relational practice, a key subject that both enables and is produced in this process is that of the host. Furthermore, that host 'must be assured of his sovereignty over the space and goods he offers or opens to the other as stranger' (Derrida 2000: 14). Without a host, the home and its ethos would also be absent and thus the ethics of hospitality would be immediately cast into doubt. A major conceptual and practical problem of conceiving refugee camps as spaces of hospitality, then, is that there is rarely a singular sovereign host with 'mastery' over the camp space. UNHCR specifies that, while it has 'unique statutory responsibility' for providing international protection, the 'provision and distribution of material assistance' can be carried out by NGOs and host governments. Indeed, 'wherever possible' assistance is provided by an agency other than UNHCR (2007: 116).

For the most part, camp spaces are governed not by a singular sovereign master, but an assemblage of different actors and authorities. In Manuel DeLanda's (2006: 3–5) terms, an assemblage is a whole constructed from heterogeneous parts. It is not reducible to those different parts but rather emerges from their very interaction, never becoming a seamless whole. Thus, 'humanitarian government does not have an actual organized global coordination, even though this is indeed imaginable' (Agier, 2011: 202). Likewise, the host of most refugee camps tends to emerge from the interaction of the various agencies, organisations and charitable bodies that produce it, secure it and govern it as a space. In Jordan's Zaatari camp, for example, which was set up in July 2012 for Syrians, a range of different actors provide food (the WFP and Save the Children), education (ten different agencies and NGOs, including Children Without Borders and UNICEF), health care (ten different actors including the International Rescue Committee and Médecins du Monde), shelter (UNHCR, UNOPS and the Norwegian Refugee Council) and community services (UNHCR and the Noor Al Hussein Foundation).[4] In camps such as Dadaab, Kenya as the 'host' state merely provides the local police, wearing UNHCR uniforms, to secure the camp and its boundaries; a vast array of different actors carry out the majority of hosting duties (Horst, 2006: 116–117; Agier, 2011: 135).

All these actors are considered 'stakeholders' in the provision and administration of refugee camp hospitality, including the displaced population themselves (Corsellis and Vitale, 2005: 30–31). No singular host controls the space; rather, its practices and ethos of hospitality are governed by an array of non-sovereign parts that interact to create a greater, non-seamless whole. Camps such as Killis in Turkey are rare, with Turkish authorities retaining almost complete control (McClelland, 2014). Likewise, the Sarahawi camps in Algeria are refugee-controlled and unusual in this opposite sense (Herz, 2013: 155). But equally, in all cases of hospitality the host is never a stable identity or subject position: it emerges in relation to, and confrontation with, the stranger and potential guest. It is this encounter with 'externality' that gives even the most apparently cohesive subjects its coherence and substance; without it the host 'would simply not exist' (Kristeva, 1991: 96). It is therefore in confrontation with the displaced population that the refugee camp 'host' is assembled from its interacting parts, coming into being rather than pre-existing the displacement in each case.

## CONTESTED ETHOS OF HUMANITARIANISM

Recognising the non-sovereign, assembled nature of the camp host nonetheless challenges us to think about precisely what *kind* of ethics, what kind of *ethos*, is practised through its hospitality. The very possibility of hospitality *as* ethics is dependent upon its describing an ethos, a way of being and dwelling that constitutes the stranger as guest and in turn is established through contact with that stranger. The ethos of a space, or home, is what gives hospitality its affective element as a welcoming of the unfamiliar, that which does not belong. This question is especially important because the refugee exemplifies the figure of the stranger in international politics (Dillon, 1999), and as such contains an inherent ambivalence between the recognisable figures of friend and enemy (Friese, 2004: 69).

For many, it is in fact the ambivalence of the international state system which is expressed spatially and ethically through the camp. These spaces are unfortunate and emergency stop-gaps for the UNHCR, which are best avoided altogether (UNHCR, 2007: 8–10, 206–208). They are not equivalent to the 'durable solutions' of repatriation, integration or resettlement. All three such solutions have the advantage of protecting, reproducing and further entrenching what Malkki (1995: 5) has called 'the national order of things', where refugees are reinstalled back into the national, state-based system of belonging that their very existence problematises. Camps only exist beyond the initial 'emergency phase' when the national order has failed: the refugees' home state is unwilling or unable to protect their people; the host state will not accept local integration; and Western states will not share the 'burden' of hospitality. Thus, the UNHCR exists to protect the norms of the international system (Soguk, 1999; Nyers, 2006), and this is equally the case for the assembled host of refugee camps. The UNHCR explicitly states this, stressing that its '[i]nternational protection is a temporary substitute for the protection normally provided by States to their nationals ... *UNHCR is not a substitute for State responsibility*' (2007: 17 – emphasis in original).

Planned camps are considered a 'last resort' (Corsellis and Vitale, 2005: 124) in part because they represent the explicit *exclusion* of refugees from the international system, the national order, and from the Western states that benefit from it. The ethos of refugee camps, the 'way of being' or dwelling which they exemplify is, on this reading, an elaborate evasion of responsibility by European, North American and Australasian states. While much work on hospitality has criticised the hostility of Western states to immigration and asylum (Dikec, 2002; Doty, 2006; Bulley, 2010) and worked towards ways to improve this situation (Carens, 2003; Benhabib, 2004; Gibney, 2004), such work can miss a geopolitical reality: 'the vast majority of refugees remain in the developing world and the refugee regime is now being utilised to contain refugees there' (Horst, 2006: 205). Camps are therefore a spatial, ethical product of this ambivalence, working to both save and contain. Like humanitarian interventions, camps for the displaced demonstrate the ethos of the Western states that fund the UNHCR, an ethos that strives *not* to come face to face with the suffering stranger (Orford, 2003: 211). As Hyndman asks, 'At what point do charitable acts of humanitarian assistance become neo-colonial technologies of control? The line is fine' (2000: 147).

While crucial to acknowledge, this reading of refugee camps and their ethos folds them *purely and simply* back into a claim about the overarching importance and centrality of states and state sovereignty. It assumes that camps are not important spaces *in themselves*, but only insofar as they tell us something about states and the international system. Indeed, this is not a reading of the ethos of refugee camps and their assembled host but of the exclusionary ethos of Western states. It chooses to ignore the everyday, mundane practices of hospitality that occur in camps in favour of the totalising claims of which IR is so fond. In doing so, it downplays the importance of these spaces, as well as the disordering and unsettling impact they can have on the international system. While it is crucial that the Western-based state system, represented by the UNHCR, produces these potentially disruptive spaces elsewhere in order to bolster their own sovereignty (Hyndman, 2000: 22–23), the tendency in critical literature has been to focus on the bolstering and ignore the disruption. This reading of camps thus works with conventional IR and international ethics to efface post-sovereign spaces in which a more banal, everyday ethics operates. It remains a critique of sovereign ethics, rather than a critical exploration of the 'changing identity spaces' that constitute an 'ethics of post-sovereignty' (Shapiro, 1994: 488).

If we pay attention to the ethos of camps themselves, then, we find great variability which evades easy generalisation. This is in large part because, as I illustrate below, once hospitality occurs the host no longer retains control over the space as its tactics are resisted by guests, and the space reoriented to new purposes. However, an effort must be made to outline the way the assembled host envisages its own ethos as it develops in relation to the stranger. To this end, the next section will examine the intertextual manifesto of hospitality produced by part of this assemblage of hosting practices. What emerges from these texts is an ethos of temporary humanitarian government, where the space of the camp is to be governed like a 'home' via mechanisms of security. This ethos makes the camp 'home' very different to the idealised, romantic view of the family homes and homeland that emerged in Chapter 1.

Given that the assembled host is one of 'humanitarian' organisations and agencies, it is unsurprising that it embraces a humanitarian ethos; 'an ethos that, for all its grandeur and moral ambition, was initially realistic enough to understand that it could aspire to do little more than alleviate suffering' (Rieff, 2002: 91). The camp host's way of being is thus inherently humanitarian, in that it privileges dignity and responds to the suffering of humans, but little more. However, 'humanitarianism is not a timeless truth but an ideology that has had particular forms at different times in the contemporary world' (Edkins, 2003b: 254), emerging from phases of colonialism and developmentalism into its present form. That present form is primarily managerial, bureaucratised and technocratic, in the sense that humanitarian agencies responded to criticisms of their practices in the 1980s and 1990s by setting out clear criteria and procedures that could be applied in a neutral manner to individual situations. Thus, humanitarianism is now characterised by a 'growing organisation and governance of activities designed to protect and improve humanity' (Barnett, 2011: 10). The latter, married with the aim of alleviating suffering, is precisely what we see in the intertextual guides and handbooks which I examine below. Humanitarianism has thus become a way of both saving *and* technocratically governing the conduct of the poor and suffering worldwide (see Duffield, 2001; Ilcan and Lacey, 2013). An ethos of humanitarianism is an ethos of 'care and control' (Malkki, 1995: 231).

Humanitarianism has thus become *a form of government* – specifically 'humanitarian government', which Didier Fassin (2012: 1) refers to as the 'deployment of moral sentiments' in the 'set of procedures established and actions conducted in order to manage, regulate, and support the existence of human beings'. Humanitarian government now operates through a variety of different governing techniques (see Ilcan and Lacey, 2013). In the hospitality provided by refugee camps those techniques can be seen as working through a form of domopolitics. As coined by William Walters (2004) and used by Jonathan Darling (2011, 2013), domopolitics refers to the tactics of governing a space as a home. While they use the term to describe the control of national space, drawing together 'an array of techniques of security designed to "secure" and regulate the place of the "homely" nation within a world of global flows' (Darling, 2011: 264), humanitarian government is not confined to states. Used in their broadest sense as moral mentalities of governing the conduct of conduct, humanitarian government also creates other kinds of spaces (Walters, 2011).

Governed spaces are produced and managed via what Foucault calls 'security mechanisms ... installed around the random element inherent in a population of living beings so as to optimize a state of life' (Foucault, 2004: 246). The key difference between state and 'post-sovereign' spaces of humanitarian government is that a state manages its *domos* primarily *for* its citizens, those who belong, who are at-home. It thus works to welcome migrants who bring an economic advantage, restricting and containing dangerous, threatening or unproductive flows. Refugee camps in contrast are produced *for* the strangers, the guests, to minimally optimise their 'state of life' and secure its continuity. Yet both produce a particular space of hospitality, with an inside and an outside, where some temporarily 'belong' (humanitarians) and others are only temporary guests (the displaced). Camps offer a less conditional hospitality than states because it is the security of the suffering guest that dominates their ethos; the space produced is thus different, though its domopolitical governmentalities are alike.

## DOMOPOLITICS, HOSPITALITY AND REFUGEE CAMPS

We can see this temporary ethos of humanitarian government emerge in the way that the assembled host plans and manages the space of the camp. In order to draw this out, I turn to the intertextual manifesto of hospitality generated by the UNHCR, Sphere Project and Oxfam/shelterproject. These handbooks set out a template for a particular kind of temporary, homely space for the guest population. This template operates in part through the three mechanisms of domopolitics that Darling (2011: 266–269) has drawn out and which I will outline in turn: the construction of a population to which power can be applied; the regulation of the mobility of that population; and ensuring the discomfort of guests to ensure their compliance. The main way in which the domopolitics of the camp operates distinctly from that of the state is in the final mechanism. While the domopolitical hospitality of the state prioritises the security of citizens, those who belong, refugee camps offer a greater priority to that of the guest. As temporary, containing, insecure spaces, camps are naturally unhomely and uncomfortable. In consequence, domopolitical humanitarianism works to establish a limited form of homeliness through community development programmes.

## PRODUCING A GUEST POPULATION

The government of behaviour via security is always exercised over a population which is characterised by contingency (Dillon, 2007: 40–42; Foucault, 2007: 11). Whereas the UK asylum system's domopolitics is based on the production of different types of subjects within a population, such as 'genuine' and 'undeserving' asylum seekers (Darling, 2011), camp hospitality is initially less fine-grained, referring to the 'disaster-affected population' (Sphere Project, 2011: 9), 'displaced population' (Corsellis and Vitale, 2005: 1) or 'refugee population' (UNHCR, 2007: 73–74) as the subject of action. Nonetheless, this specific population is not a naturally occurring entity; it needs to be created. And this has to be done in advance, to produce the particularities of the camp space before the guest population's arrival.

What Oxfam/shelterproject call the initial 'preparedness' and 'contingency' phases of an operation are therefore largely about the collection of information on the population to be welcomed and secured (Corsellis and Vitale, 2005: 40–42),[5] thus bringing this population into being. Here, it is the range of population statistics and their contingency, their uncertain movement, which actually generates the emergency that requires the spatial technology of a camp in the first place. Emergency indicators engender crisis through measurement, by far the most important of which is the mortality rate. If this is greater than two deaths per 10,000 of the population, per day, an emergency exists and camps may be required. Another key indicator is child nutrition – an emergency exists if there is '10% with less than 80% weight for height' (UNHCR, 2007: 64 and 546). The technicalities of this humanitarian biopolitics are most apparent than in the 'Z-score' for determining malnutrition; the UNHCR's 'preferred mode of presenting anthropometric indicators in nutrition surveys'.[6]

Once an emergency is established through biopolitical security mechanisms, we enter the 'transit' phase, allowing 'influx management' techniques through which a greater population profile is built (Corsellis and Vitale, 2005: 43). This demands 'regular updating of sufficient, accurate and timely information on the size, position, and composition of the influx of displaced people' (ibid.: 357). A humanitarian welcome is thus not directed towards an unknown stranger, and neither is it a singular other; rather, it targets a mass, thoroughly known through its statistical composition. For this reason, each way-station, transit centre and reception centre along the route to the main camp is equipped with at least a preliminary registration capacity. Each such space welcomes on the basis of the humanitarian ethos of care *and* control (Malkki, 1995: 231), offering rations, water, and a health screening, but also counting refugees to determine 'influx rates'. Without these intermediary spaces, camps cannot offer the security as optimisation of life that their various hospitable tactics are directed toward. Each facility keeps refugees alive while forming them as a 'displaced population' – a known, calculated collection. In fact, this is a never-ending construction exercise; registration is a 'continuous process' (UNHCR, 2007: 158). Registration operates through four phases: determining the strategy; collecting information and distributing cards; computerisation using *proGres* software; finally 'verification and updating' (ibid.: 162–165). The latter is endless by its very nature, as the population is ever-changing, depending on the development of the crisis or emergency, which is itself determined in relation to that shifting population.

## REGULATING MOBILITY

Like the production of the guest population, the attempt to map and govern its spatial mobility begins before a crisis even occurs and continues whilst the displaced are 'in transit': the 'contingency phase' of an operation is the 'period before an emergency which is yet to occur but is likely to happen' (Corsellis and Vitalle, 2005: 41). While the 'displaced population' are in process of being displaced, the 'influx management' phase then governs the 'process of supporting and guiding the transit of displaced populations away from danger, and towards appropriate TS [transitional settlement] options' (ibid.: 357). If this mobility is correctly and pre-emptively conducted, it allows for a better 'reception' and securing of the displaced by ensuring sufficient capacity is available on site. But this means controlling the 'influx' (i.e. the movement of the displaced) before it even begins. To facilitate this, a 'network of support and pre-registration facilities' are erected, as mentioned above, consisting of 'way-stations, transit centres, and reception centres', each with sufficient protection, capacity and clean water (ibid.: 258). A fictitious example allows this situation to be 'mapped' in terms of how each facility stands in relation to the others (ibid.: 359). Here, the broad 'direction of flight' is identified by a large arrow and various paths designate where movement is safest as well as the distances involved (half a day's walk between each facility). The 'movements of vehicles and people should be organised into an efficient system' at each transit point along the route (ibid.: 359).

The assembled camp host is thus exercising a form of pastoral power similar to that which we saw in *Welcome to Sarajevo* (Chapter 1). Such a 'beneficent' power of care operates over the 'flock' of refugees as a 'multiplicity in movement' (Foucault, 2007: 125–128). Effectively, camps as spaces of hospitality produced through the regulation of mobility are extended spatially and temporally far beyond their apparent boundaries of welcome and containment. They arguably extend to the very moment and place of displacement, with transit centres forming *part of* the camp's securitised hospitality. Each facility along the line of flight is individually mapped out (Corsellis and Vitale, 2005: 363–366), with areas for assembly, registration, health screening, distribution centres, food preparation, latrines, accommodation and departure. Minimum spatial standards are specified, with reception and transit camps requiring at least $3m^2$ per person (though Sphere (2011: 259) suggests $3.5m^2$); at least $100m^2$ per 500 people for food preparation (Corsellis and Vitale, 2005: 361) and at least 100m from accommodation to refuse disposal (ibid.: 366; UNHCR, 2007: 223). These maps are organised around small arrows directing the movement of refugees through each facility, from their guarded entry to their guarded departure. The aim is for quick, efficient and safe circulation, allowing the 'influx' to be managed effectively.

Abundant advice is offered on the spatial planning and organisation of the final (though temporary) destination camps. The creation of a 'master plan' is essential (Corsellis and Vitale, 2005: 368; UNHCR, 2007: 206). This is an overall site plan, regularly updated, marking boundaries, sub-divisions, infrastructure and facilities (Corsellis and Vitale, 2005: 382), but also mapping the social organisation of the refugees alongside topographical and 'planimetric' surveys (UNHCR, 2007: 215). This allows for continuity of control despite the rapid turnover of international staff. It also allows the space to continually shift and change with the contingent guest population. A range of minimum standards are outlined, including the location and number of tap stands, latrines, showers, fire breaks, distance between buildings and blocks, number and location of refuse disposal sites (Corsellis and Vitale, 2005: 278–279; UNHCR, 2007: 206–211; Sphere Project, 2011: 239–259). The headline figure is that the minimum surface area is $45m^2$ per person, including infrastructure and household agricultural plots (Sphere Project, 2011: 257; UNHCR, 2007: 210). This creates a mathematics of hospitality, where the size of the space is determined by the numbers to be secured. Thus, 20,000 people – and all advice is to discourage camps larger than this to avoid an unmanageable population (UNHCR, 2007: 211; Corsellis and Vitale, 2005: 371) – generates a space as: '*20,000 people x $45m^2$ = $900,000m^2$ = 90 hectares (for example, a site measuring 900m x 1000m)*' (UNHCR, 2007: 211, original italics).

What is not shown in these guidelines because they are abstracted from concrete examples is the way that, once established, the camp space works to control and limit movement. The UNHCR specifies that IDPs, as citizens in their own country, 'should not be forcibly restricted to "camps" and they should have the freedom to move in and out of camps' (UNHCR, 2007: 32). No such requirement exists for refugees whose non-belonging automatically removes any such freedom. Thus, from the transit and reception centres onwards, Agier (2011: 150) notes that refugees are 'processed' through categorisation procedures. In multi-ethnic camps such as Maheba in Zambia, refugees must wait in squalid and overcrowded conditions after registration before being taken to the 'zone' in which they are thought to belong (2011: 120–126). Even without zonal constraints, there is often highly restricted movement beyond a camp's boundaries. While Palestinians in Lebanon have experienced both relative freedom (from 1968–75) and stark confinement (from 1982–95) in different periods (see Peteet, 2005: 6–11), the remoteness and isolation of camps such as Mishamo in Tanzania and Dadaab in Kenya has always severely restricted mobility. Indeed, 93 per cent of camps world-wide are located in rural areas, with few transportation options (UNHCR, 2013: 37). Remoteness is sometimes an inevitable consequence of the requirement to minimise 'negative impacts on the local host population' in selecting a site (Corsellis and Vitale, 2005: 355). Nevertheless, even in isolated Mishamo and Dadaab, movement is also regulated using leave passes (Malkki, 1995: 138) and identity documents (Horst, 2006: 23).

## GENERATING (AND TEMPERING) COMFORT

The Sphere Handbook (2011: 244) explains that, while shelter is crucial for survival and safety, '[i]t is also important for human dignity, to sustain family and community life and to enable affected populations to recover from the impact of disaster'. Humanitarian hospitality thus increasingly aims to provide more than survival, including recovery, dignity and the sustenance of 'goods' like family and community. Camps aim to offer 'an entire support system' (Corsellis and Vitale, 2005: 115). The UNHCR emphasises a major change in protection over recent decades, with refugees necessarily 'involved at the heart of decision-making concerning their protection and well-being' (2007: 82). Its 'community development approach' thus sees all displaced individuals as 'resourceful and active partners' in their own protection (ibid.: 182). This includes consultation with refugees on the siting of camps (Corsellis and Vitale, 2005: 127–128), the provision of goods and education (UNHCR, 2007: 415). The aim of community programmes is explicitly constructive: traditional structures may have 'broken down', so programmes like 'participatory assessment' help to 'mobilize communities' (ibid.: 182–183). In contrast to the domopolitics governing the state's treatment of asylum seekers (Darling, 2011), humanitarian domopolitics aims to manufacture comfort and 'homeliness' in a situation of threat, violence and insecurity. The host tries to make the guest feel 'at home' through a concentration on family and community, seeking to reconstruct a minimal comfort and sense of temporary belonging for guests.

We can see this reconstructive effort in the suggested materiality of the camp space. The UNHCR specifies that the 'basic planning unit' for the organisation of camps is the family, and 'larger units' then 'follow the community structure' (2007: 106, 186). The camp should be organised around 'the smallest module, the family' and then build up through a community (approximately 16 families, or 80–100 people), followed by a 'block' (16 communities), a 'sector' (4 blocks), and the camp as a whole (4 sectors) (ibid.: 216). Site-planning should thus always be 'bottom up', reflecting the wishes of the community and its need for infrastructure, rather than the desires of administrators (ibid.: 213). Each family plot includes room for a kitchen garden, enabling a move towards 'self-sufficiency' and 'independence'. Interaction between groups is actively encouraged through spatial design, suggesting that communities 'are not closed form, e.g. square shaped, but resembling more of an H-shape, where both sides are open for better interaction with other communities' (ibid.: 213). Taking all this into account, then, the model camp is organised with a 'community' at its base in *Transitional Settlements* (Corsellis and Vitalle, 2005) – see Figure 2.1. In this map, each of the rectangles represents a family plot with its own latrine; each black square a family dwelling; each 'x' a tap stand; and each circle a shower. Far from making community impossible (see Hyndman, 2000: 137–145), the idealised spatial structure of camps can go to great lengths to cultivate it, even with the restrictions on mobility it simultaneously enacts.

As I have argued elsewhere (Bulley, 2014a), this production of the camp as a welcoming, homely, communal space is both laudable and riven with power relations. Community building and encouraging the participation of the displaced in the construction and siting of the camp space also allows greater control to be exerted over the guest population. For Ilcan and Lacey such 'community-targeted empowerment' (2011: 14) uses 'community' as another 'collective label' (like 'villagers' or 'the rural poor') allowing for more effective government (ibid.: 26). Inevitably, the UNHCR's 'actions' surrounding community empowerment include gathering information on control of resources and decision-making structures in order to '[s]ystematize the information to build a picture of the population profile' (UNHCR, 2007: 185; see also 58).

However, more than just a collective label, community is used to shift responsibility towards refugees themselves, to make them part of their own security management – the aim of mobilising communities is explicitly so they can 'enhance *their own* protection' (UNHCR, 2007: 183). This is evident in the provision of kitchen gardens to allow families 'self-sufficiency' and 'independence'. Likewise, community involvement in 'commodity distribution' (ibid.: 229), 'food-for-work' schemes (Corsellis and Vitale, 2005: 192) and 'self-help labour' are encouraged in community construction projects as they offer choice, 'skills training' and 'broaden livelihood opportunities' (ibid.: 195). This reflects changes in the way arenas such as welfare and health have been governed in industrialised states around the world since the 1980s (see Miller and Rose, 2008).

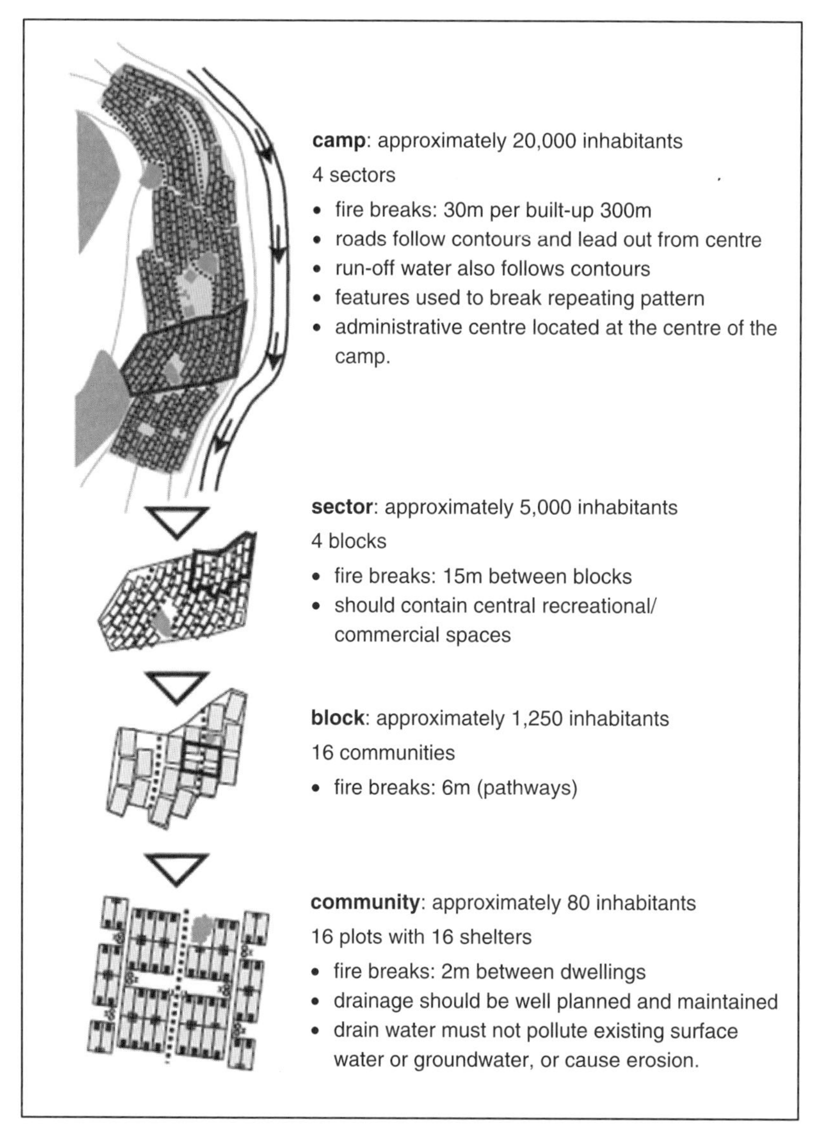


**Figure 2.1** The community and the camp (reproduced from Figure 8(e), Corsellis and Vitale, 2005: 380)

Effectively, the aim is to make 'the poor', wherever they are located, responsible for their own relief and less dependent on regular aid (Lippert, 1999: 313).

In this sense, the assembled host uses a particular and contestable concept of community and comfort in order to exert more subtle forms of control over the hospitable space. The displaced guest is no longer just another 'stakeholder' in their own protection (Corsellis and Vitale, 2005: 30–31) – while making the camp more comfortable and homely, their constitution as a 'community' (within a population) also provides opportunities for their management. They become a subject of more indirect intervention. Traditional power structures within the displaced community can be utilised, but also constantly interfered with (as they are 'not necessarily fair' – UNHCR, 2007: 106, 186) to suit the securitising humanitarian ethos of the UNHCR and NGOs and their gender and minority rights concerns. Community capacities, resources and labour can be exploited to a greater degree, while 'empowering' refugees at the same time (Corsellis and Vitale, 2005: 195). Community 'participation' and 'involvement' becomes a matter of changing behaviour and habits on issues such as personal hygiene by 'providing information to the community on life in their new situation, which may be markedly different from their previous experience' (ibid.: 106–107). Producing, encouraging and 'empowering' community is thus enfolded within wider forms of governmentality which aim to secure life and thus impinge on every element of the guest's existence within the camp.

The humanitarian domopolitics of camp hospitality concentrates on producing 'homeliness' but also maintains elements of discomfort, unease and insecurity. Thus Oxfam/shelterproject warn against the 'over-provision of support' in camps, as this potentially leads to resentment among local populations, the displacement of even more people (the 'honey pot' effect), even attracting elements of the host population, the destabilisation of local economies and the development of dependency (see Corsellis and Vitale, 2005: 15). This produces the 'principle' that service-provision within camps 'should not be higher than that available to the local population' (UNHCR, 2007: 122). There is also a stress throughout the intertextual manifesto on the limited temporality of humanitarian hospitality. Camps are merely a 'temporary substitute' to state-based protection (ibid.: 17) and should not become a 'default response' (Sphere Project, 2011: 251) or 'default permanent housing' (ibid.: 259). Camps are always *transitional* settlements (Corsellis and Vitale, 2005).

Ensuring temporariness and the absence of 'over-provision' works to constantly temper the homeliness of camp spaces and ensures they exist as in-between places. They are based on land ceded or leased by the host state to the temporary jurisdiction of the international community, generally represented by the UNHCR (see UNHCR, 2011: 121, 221) or the United Nations Relief and Works Agency (UNRWA). Such territory may be taken back at any time. This precarious insecurity is illustrated through the absence of camps from official maps of state territory: though the world's largest refugee camp, the three sites at Dadaab, built almost two decades ago, covering a radius of 15 kilometres and being densely populated by nearly 500,000 refugees in 2011, do not appear on authorised maps of Kenya (Agier, 2011: 135). The sites at Dadaab are often a reference point for the insecurity of camps, especially after a Kenyan Parliamentary Committee announced plans to close these spaces in the aftermath of the Wakefield mall attack in 2013 (see BBC, 2013).

Though impermanent, many camps also cannot adequately be described as 'temporary'. For example, the Sahrawi camps in Algeria are over 35 years old and increasingly well-entrenched (Herz, 2013). Over half of the nearly 8 million Palestinian people are registered refugees, with the majority of those living in Lebanon based in camps which have lasted for over 50 years (Peteet, 2005: 6). Nonetheless, this leaves all camps as limbo spaces, neither permanent nor temporary, potentially homely but not too homely. This ensures that these spaces of post-sovereign hospitality are never too disruptive to the national order or the international system that sustains it. As spaces of security and existential insecurity, the incentive always remains to seek and seize any opportunity for the 'durable solutions' of returning to a state-based home. And of course, this degree of unhomeliness also ensures that refugees remain compliant with the limited hospitality and control mechanisms of the camp space.

## COUNTER-HOSPITALITIES: SEIZING AND REDIRECTING CAMP SPACES

Taking refugee camps seriously as post-sovereign spaces reveals the extent to which they rarely conform either to their generalised representation in much academic literature, or the technocratic plans of the internationally assembled host. Far from a blank space waiting to be filled with bare life, camps are a *product* of hospitality and the interaction (and blurring of boundaries) between hosts and guests. As Ramadan (2013: 70) notes, a camp is 'the people within it and the relations between them: the space and the society are one formation, a camp-society ... not a monolithic body with a single pure identity, but a diverse, dynamic and at times divided assemblage in constant motion'. Camps are thus profoundly political spaces in which hosting power is exercised to conduct guest conduct, but is also constantly countered, its directives redirected. The Foucauldian notion of counter-conduct is useful here because the actions that work to subvert hospitality are not necessarily intentionally political or revolutionary; rather, they are often ordinary, banal, 'diffuse and subdued forms of resistance' (Foucault, 2007: 200–202). Yet they can work toward seizing the home, its space, materiality and ethos. Thus, '[e]ven when they [refugees] have not engaged in overt politics, the engagement with spatial practices themselves has not only been subtle critiques of humanitarian assistance and states of exception, but also discreet ways of subverting the condition of liminality' (Sanyal, 2011: 880).

These everyday forms of resistance cannot be discerned from the host's intertextual manifesto examined above, as this does not deal with or anticipate the failures and subversions of humanitarian hospitality. Instead, we must turn to the work of ethnographers, geographers, architects and sociologists who examine precisely the way displaced people live and transform spaces of temporary hospitality. Here we can see the way in which, because the camp as a space is a 'sphere of coexistence of a multiplicity of trajectories' (Massey, 2005: 63), the assembled host can never control those trajectories, the coexistence, or what it will produce. This final section of the chapter examines some ways in which *specific* camps have been and are being spatially transformed, host and guest are swapping positions, allowing a counter-ethos rejecting humanitarian government to emerge.

## GUEST IDENTITY GAMES

Many of the counter-conducts practised by displaced people involve disrupting the attempt to fix their identity as a population. These are not undertaken, however, *as* a population, but as individuals, families and networks of communities, undermining the humanitarian-domopolitical government of hospitality and its need to 'fix' the population of a camp (UNHCR, 2007: 163). Common strategies used by refugees to secure more rations include the double-entering of names on the host's lists, registering in more than one zone/village, adding fictional family members and non-registration of deaths and departures (Peteet, 2005: 72). More subversive are the identity and ration card 'games' played by Somali refugees and Kenyan locals in and around the three camp sites in Dadaab. For example, when individuals and families depart, it is common for them to leave their ration cards with family or friends (Horst, 2006: 94). These can then be used to increase the rations received by the remaining family, while any surplus can be traded. The card itself can be traded at local markets for money, goods and services. Because of the access it provides to food and amenities, ration cards are precious commodities, sought-after by less affluent local Kenyans. Due to Dadaab's location close to the Somali border, it is often difficult for Western authorities in the camp to distinguish between Somalis and Kenyans, let alone between refugees and non-refugees (Horst, 2006: 23). One UNHCR official claimed that of the 45,000 claiming rations in Ifo camp, around 11,000 were local Kenyans (Horst, 2006: 95). Such manoeuvrings need not prove to be moments of 'solidarity' between refugee and local populations (Millner, 2011); rather, they are counter-conducts produced by the interactions of different identities and their trajectories that camps as spaces allow.

Manipulation of control through such games can also increase wider mobility and the options available to refugees. A particularly prized exchange for a Dadaab ration card would be a Kenyan identity card, which allows the possibility of travel throughout Kenya denied to refugees. Identity cards, driving licences and work permits can also be bought and sold with the complicity of Kenyan officials (Agier, 2011: 138), blurring the identity of the guest population further and undermining the regulation of its mobility. In a similar way, Malkki outlines the pragmatic means that Burundian Hutu 'town refugees' in the Kigoma region of Tanzania found of manipulating their identity – changing religion, inter-marriage, vague identity descriptors – in order to become invisible and thereby increase their freedom of movement outside the camp (Malkki, 1995: 153–183).

Even where the camp is successful in fixing the identity of the guests, this can have unintended consequences which generate disorder rather than greater control. Malkki thus notes the stark opposition between the culture and identity politics of 'town refugees' in Kigoma to the 'camp refugees' in Mishamo. Tanzanian authorities had set up Mishamo to cope with the influx of Hutu refugees fleeing genocide in Burundi during 1972–73, although it was later taken over by the UNHCR and its assembled hosting practices. Despite the fairly severe restrictions exerted over the guest population in and through Mishamo, it 'unexpectedly' became 'a site that was enabling and nurturing of an elaborate and self-conscious historicity among its refugee inhabitants' (Malkki, 1995: 52–53). Far from removing their identity, the space of the camp and the way it concentrated, confined and restricted mobility enabled the development of a 'mythico-history' of the Hutu people. This acted as a 'charter and blueprint' through which to interpret their exile and all the actors, past and present, that had contributed to it: the 'evil' Burundian Tutsis and Tanzanian authorities; the 'benign', if gullible, Belgian colonists, UNHCR officials and NGO workers; and even the 'town refugees' as 'impure' Hutus (ibid.: 144–153). Instead of making the guests more amenable to humanitarian government, this distillation of identity generated by the camp made its refugees suspicious and resistive of attempts by the assembled host to regulate their conduct through agriculture, taxation, education and vaccination (ibid.: 122–130).

Even when the production and regulation of a guest population is seemingly successful, the consequences of the identity generated can therefore be the opposite of its intention. An excellent example is Lebanese refugee camps, established in the late 1940s and 1950s after the violent displacement of Palestinians following the creation of the state of Israel. UNRWA, set up to manage this population, was especially active in the sphere of education, organising free elementary and middle schools and training centres. This was part of a long-term governmental project to 'enable refugees to support themselves and form a mobile regional labor force' throughout the Middle East, allowing for their 'rehabilitation and resettlement' (Peteet, 2005: 86). Formalised education drew families and communities into the camps (ibid.: 87). While the refugees had no control over the curriculum, the education and employment of Palestinian teachers allowed Palestinian history and geography to be smuggled into lessons (ibid.: 90). Such everyday counter-conducts paved the way for more spectacular forms of resistance. Though an educated workforce was formed, the unforeseen impact was that secular education provided 'the means to transmit a Palestinian national identity ... prepar[ing] a generation of educated youth for secular, militant nationalist activities'. The result was 'the emergence of the resistance movement in the camps' (ibid.: 88–89). The guest population that was to be durably resettled following their governmental moulding in the camps, redirected these techniques in the service of proliferating a nationalist identity. This produced violent resistance to displacement and a semi-permanent residence in these 'temporary', containing camp spaces. Instead of durable solutions, concentration and education helped extend the temporality of humanitarian hospitality indefinitely.

## BUILDING MEANING, HOMELINESS AND ETHOS

A key element governing the ethos of refugee camps is a limited temporality; their ethos is one of *temporary* humanitarian government. Yet as the immediate 'crisis' and 'emergency' often stretches into years and decades, displaced people inevitably begin to counter the space's ethos, grafting on their own, implanting different habitual ways of being and dwelling. It should be noted that some displaced people find this easier than others. Horst (2006) argues that Somalis have a long history of living with insecurities and have developed a series of coping mechanisms. This 'nomadic heritage' includes frequent movement to greener pastures, strong social networks that entail obligations of assistance and the spreading of investments to reduce risk. These mechanisms have not only survived in Dadaab but have 'now acquired a transnational character' (ibid.: 2–3). Social networks are 'complex human relations in which individuals negotiate and at times switch (power) positions, in order to survive' (ibid.: 237). They function on the basis of transactions that spread information, services and resources, beginning with family but encompassing clan-members, neighbours and the proximate needy. Coalitions based on networks are neither stable nor unitary but continually being built and rebuilt (ibid.: 63–70). The transnational character of this ethos has been aided by the introduction of electronic and radio technologies into Dadaab in 1998, helping produce a vast increase in the flow of information and remittance money into the camps (ibid.: 130–157).

In other situations, especially when the displaced have no 'nomadic heritage', a counter-ethos has been gradually built through the grafting of meaning and alternative geographies onto the camp spaces. Examples can be found in studies of Palestinian refugee camps in Lebanon (Peteet, 2005; Ramadan, 2010, 2013; Sanyal, 2011). Peteet's detailed study in particular shows that initially meaning was given to these blank spaces through their organisation according to villages of origin. Refugees imposed 'their own sense of spatiality on the camps, crafting a microcosm of Galilee'; four generations on from displacement, Peteet found children defining themselves as being from villages in Galillee they had never seen (2005: 100–101). During this period the necessity of identity cards and checkpoints were also productive of lines of inclusion and exclusion which starkly reinforced the 'placemaking' of the refugees (ibid.: 128).

Villages thus provided a point of departure for placemaking, but were gradually 'overlaid with other sorts of space', from the institutional (distribution centres and schools) to informal social gatherings (*dawaween*), inter-village marriage venues, and the markets and bakeries which produced an embryonic national cuisine (ibid.: 116–117). Agier found that similar spaces materialised in Dadaab, with the emergence of markets, Ethiopian coffee shops and the Ethiopian and Sudanese 'video shops' showing Indian films (2011: 134–137). This repurposing of space and materiality is not necessarily overtly political, but nonetheless builds alternative ways of being and dwelling, outside or alongside the temporary humanitarianism of the assembled host. In particular, the importance of material 'things' cannot be underestimated as they are invested with meanings, memories and affective feelings of 'belonging'. The materiality of a culture and ethos matters. Objects in a dwelling are rarely 'a random collection. They have been gradually accumulated as an expression of that person or household' (Miller, 2008: 2). They help produce the home and feelings of being at-home.

The notion of 'building' meaning is therefore not purely metaphorical. While the Palestinian refugees in Lebanon were initially housed in UNRWA tents, by 1961 these had all but disappeared (Peteet, 2005: 109). Sanyal describes how this happened covertly because of the ban on building with solid materials. Adobe bricks were made from sand or clay before being used to build walls inside the tent – in a perfect metaphor, her interviewees noted that 'from the outside, all that could be seen was the tent, yet on the inside there was a solid structure being constructed' (Sanyal, 2011: 883). Food tins could then be beaten into sheets to form further walls and ceilings. While not revolutionary, such counter-conducts were about the 'making and unmaking of self and space ... camouflaging their resistance and rescripting their identity' and ethos in response to the assembled host's attempt to impose a temporary way of being.

Once tents are replaced with solid dwellings, further layers of meaning can be added to these material structures, generating a homeliness beyond that prescribed by the host. In the Lebanese camps this was done through readopting the Palestinian chequered headscarf (the *kuffiyeh*) and producing communal arts and crafts incorporating Palestinian flags and maps in jewellery and home decor (Peteet, 2005: 149–150; Ramadan, 2013: 73). In a similar way, Western Saharan refugees in Algeria have incorporated the Sahrawi flag into house designs and furnishings (Herz, 2013: 13). In the Palestinian camps however, as the 1990s brought a much greater violence and suppression of their identity and freedom, these flags, cultural symbols and maps fell away somewhat. Homes came to be decorated by photos of the dead (Peteet, 2005: 209–212) as nationalist layers of meaning ceded to anguished memories and 'layers of trauma' (ibid.: 206). Though traumatic, such layers also contributed to the construction of the space, its meaning, and the way of *being* a refugee that both constitutes and is constituted by humanitarian hospitality. For Palestinians, it helped generate and sustain an ethos of resistance to both their displacement and their assembled humanitarian host, even after the Israeli invasion in 1975, the 15-year civil war and the repression of the 1990s. In turn, the militarised resistance in the camps from 1969 allowed greater freedom to embellish the home, making it more liveable by building concrete roofs, second storeys and private bathrooms (ibid.: 133). As noted earlier, while a population was formed, mobility was restricted and feelings of community were generated, this did not lead to more easily controlled guests.

Yet these counter-conducts did not subvert all aspects of their control; elements of the governing tactics that make camps *temporary* spaces were supported. Few Palestinians wanted to make their guest-condition permanent; most longed for a return to the 'national order' that is central to the temporary ethos of the camp. This leads to a contested politics of homeliness that incorporates the tempering of belonging envisaged by the assembled humanitarian host. For example, the construction of solid, permanent housing in camps is frowned upon and treated suspiciously by many Sarhawis who retain a preference for the impermanence of tents. These both symbolically and literally demonstrate readiness to return home to a liberated Western Sahara (Herz, 2013: 112–150). Likewise, the houses built by Palestinians in Lebanon were intended to be left to the Lebanese – 'As self-perceived guests and seekers of refuge in Lebanon, they thought to reciprocate long-term hospitality' (Peteet, 2005: 133). A key spatial signifier in the camps were 'prominently placed markers specifying their distance from Palestine, the homeland, keeping alive the hope of returning' (Sanyal, 2014: 568). Thus, as Foucault notes, counter-conducts operate as 'border-elements', not entirely external to governmental strategies (Foucault, 2007: 214–215); they reinforce and bolster the dominant form of government, as well as undermining it (Death, 2010: 240). Refugees in these post-sovereign spaces rarely resist the 'national order of things' as such, but demand a return to it *on their own terms*.

## HOSTING AND GUESTING

While elements of the assembled host's ethos may be embraced, and its power enhanced, the emergence of a rival guest ethos supplements the host's in order to displace it. The hospitality offered by camps in each individual situation is thus a highly idiomatic work of contestation and negotiation between host and guest, between restrictions and their rejection, conducts and counter-conducts. Examining camps as spaces of hospitality offers a frame within which to view these shifting power relations and their ethical repercussions. The arrangement between guests and hosts vary dramatically from situations where the displaced have near autonomy, as in the case of the Western Saharan camps in Algeria where the Sarhawi have almost become their own hosts (Herz, 2013), to situations where a host state closely regulates the way temporary guests are welcomed and provided for, such as the Turkish camps for displaced Syrians (Dinçer et al., 2013: 12–18; Özden, 2013: 5–7).[7] Frequently, the relationship between hosts and guests is more nuanced and mobile.

One useful example of this is offered by Agier in his study of Maheba in Zambia (2011: 116–131). The land for this camp was originally ceded by local chiefs to the Zambian state, and then on to the UNHCR. During Agier's study in early 2002, three generations of refugees from conflicts in Angola, Congo (DRC), Rwanda and Burundi, encompassing many different ethnic groups, comprised the camp's 58,000 population. Each 'zone' of the camp, from A (opened in 1971) to H (opened in 1999), was created to welcome and accommodate new waves of guests, with each group settling and achieving a degree of self-sufficiency. However, new arrivals find that the assembled host of the UNHCR and NGOs 'do not take full responsibility for them', leaving their hospitality to be 'negotiated with the established refugees' (Agier, 2011: 128). Those predominantly Angolan refugees who have been at Maheba for 20–30 years have taken control of the best land, agricultural plots and jobs, working in the reception of new arrivals and the distribution of aid (ibid.: 126–127). What emerges is a highly conditional form of hospitality where established guests have effectively fused with the assembling host. The best example emerged when the camp administrators sought to open a new 'village' in March 2002 on land that established refugees were using for agriculture. A section of this land thus had to be 'released' to welcome the new guests, not by the host, but by other refugee guest-hosts. These new arrivals were then employed by the guest-hosts to work the remaining land around their village (ibid.: 128–129). Surprisingly perhaps, Agier found that many established refugees in Maheba were ambivalent or uninterested in returning 'home' to be dominated by Tutsis or their civil war enemies. The subject positions of 'host' and 'guest' and the spaces of 'home' and 'camp' are thus revealed as unfixed, mobile outcomes of hospitality practices.

Further evidence can be found by returning to the experience of Palestinian refugees in Lebanese camps. Here, the displaced population have been contained and controlled guests during their initial exile (1948–68), free and autonomous guest-hosts with the emergence of militant resistance in the camps (1968–75) and starkly confined prisoners (1982–95). However, while their status has always been one of negotiating with, and more-or-less camouflaged resistance of, the assembled host, Ramadan (2008) describes how the positions of host and guest were utterly reversed in 2006. During the summer of that year, Israel launched airstrikes and ground incursions into Southern Lebanon, targeting Hezbollah but also destroying infrastructure (bridges, hospitals, airports, sewage treatment and electrical facilities) and 30,000 Lebanese civilian homes. Over 900,000 people were displaced by the war: 'It was in this context that Palestinian refugees offered refuge and hospitality to displaced citizens of Lebanon', with the Palestinian camps in south Lebanon being comparatively safe havens for 10,000 people from the cities of Saida and Sur (Ramadan, 2008: 662–663).

Palestinians had no formal or juridical sovereignty over the camp spaces but were not forced into acts of hospitality – the 'guests' had the power to welcome or reject their nominal 'hosts'. They chose to offer more than bare assistance, welcoming the displaced Lebanese with shelter, often in Palestinian homes, medical aid and ample food, while putting themselves at risk of airstrikes (Ramadan, 2008: 667). This hospitality, as understood by the Palestinians interviewed by Ramadan, was both an ethical response to suffering and a political attempt to change Lebanese perceptions of Palestinians and their camps. It was generous *and* instrumental, while also poorly coordinated and contested by some who felt it excessive (ibid.: 669–672). This hospitable relation was reversed again after the war and had limited impact in terms of changing perceptions (Ramadan, 2008: 674, 2009: 159–163). Nonetheless, the episode demonstrates the irruptive potential of camps as post-sovereign international spaces of ethics and power relations. The counter-conducts enacted within these temporary spaces of hospitality, including the building of homeliness and a resistive ethos, were used to (at least temporarily) redefine the relationship with the other. The nominal host became the temporary guest and vice versa, revealing that a counter-ethos generated in a space born of violent and containing hospitality need not only be dangerous and threatening. It can also be nurturing, compassionate and more generous than the humanitarian government it mirrors.

## CONCLUSION

Arguing that refugee camps should be considered post-sovereign spaces created through practices of humanitarian hospitality is not to romanticise them. Refugee camps often remain places of deep and abiding insecurity. Especially early in a crisis, most refugees are entirely dependent upon the basic sustenance provided by the WFP, whose provisions often fall short of the necessary daily calorific intake leading to malnutrition and illness (Abdi, 2005: 8). Yet, as a Somali Refugee in Kakuma, Kenya, observed: ‘it is of no advantage for us to get a full ration from UNHCR if our lives are at risk from insecurity’ (Crisp, 2000: 602). Factors which produce such insecurity include under-reported domestic and community violence, both of which tend to be focused on women and children; sexual abuse, violence and rape perpetrated by those within the camp and those from outside; armed robbery and banditry; so-called ‘ethnic’ violence within and between national refugee groups; and violence between refugees and the local population (ibid.: 603–611). Women are often made particularly vulnerable by camp spaces, with similar trends of rape and forced marriage emerging in Syrian refugee camps according to the International Rescue Committee (2013: 2).

Yet this insecurity should not prevent us taking refugee camps seriously as spaces of international ethics. If, as this book demands, we are to explore an ethics of post-sovereignty then we must look to just such irruptive post-sovereign spaces where hospitality as ethical relation is ongoing, part of everyday practice. While such practices are clearly not hospitality in any unconditional sense, they gesture towards something seemingly unconditional; though the humanitarian ethos is increasingly governmental and bureaucratised, it retains a reference to something 'transcendental' (Barnett, 2011: 10), even if simply an affective 'humaneness' beyond the mere existence of 'mankind' (Fassin, 2012: 2). Yet the governmentalisation of this hospitality is key to the domopolitical construction of the camp space and attempts to control the guest population, even before its formal welcome, while refugees are 'in transit' to the official camp site. This humanitarian hospitality also targets the ethos of communities within camps in order to better manage their guests' behaviour.

However, governmental tactics are based on freedom rather than restriction alone (Foucault, 2008: 63; Walters, 2012: 31), and a range of counter-conducting hospitalities result from these tactics: resistive practices that seize and redirect the ethos and space of the camp, potentially transforming the host–guest relation. These counter-hospitalities help form part of the ethics of post-sovereignty which shows glimpses, as in the Lebanese example above, of what *could* be possible, *perhaps*, for a future of humanitarian hospitality. However, such resistances also frequently reinforce that which they apparently resist, refusing to challenge the national, sovereign state-based order to which refugees generally wish to return. What they challenge and *critique* is the need to return to the national order '*like that*, by that, in the name of those principles, with such and such an objective in mind and by means of those procedures' (Foucault, 1997: 44). In the next chapter, I will turn to a space of hospitality that emphasises an ethos of freedom and flourishing: the urban hospitality of global cities.

## NOTES

1 This rose from an average of 32,200 people per day in 2013 (UNHCR, 2014b: 2), and represents a near quadrupling since 2010 (UNHCR, 2015: 2). It is important to remember that UNHCR figures are commonly thought to be conservative (Agier, 2011: 19).

2 The Sphere Project was set up by NGOs and the Red Cross and Red Crescent Movement in 1997 to 'develop a set of universal minimum standards in core areas of humanitarian response' (2011: ii). The result is the Handbook, the third edition of which I refer to here.

3 *Transitional Settlement* is a set of 'guidelines' for agencies 'concerned with the transitional settlement needs of displaced people and their hosts' (Corsellis and Vitale, 2005: 1). This guide is the result of a collaboration between Oxfam and shelterproject, funded by the UK Department for International Development, using materials from NGOs such as Engineers Without Borders and Architectes Sans Frontières (ibid.: x–xi). Through frequent cross-referencing of the Sphere and UNHCR Handbooks (e.g. ibid.: 13–14), *Transitional Settlement* forms an exemplary piece of inter-agency governmental planning, involving all 'stakeholders' in the organisation and management of the camp space (ibid.: 30–31). The guidelines are an example of how advanced liberal management tends to 'govern "at a distance" ... by establishing networks, links, or partnerships with state and non-state actors' (Ilcan and Lacey, 2011: 51–52).

4 The UNHCR website has a useful tool for illustrating the governmentalisation of hosting practices which produce the assembled host. For each camp in the Syrian crisis, there is a side tab addressing 'Who's doing what where?', offering a clickable list of actors under the headings of camp management; child protection; community services; coordination; core relief items; education; food security; gender-based violence; health; mental health and psychological support; nutrition; protection; registration; reproductive health; shelter; and water and sanitation. For Zaatari Camp, Jordan, see http://data.unhcr.org/syrianrefugees/settlement.php?id=176®ion=77&country=107 (accessed 24 July 2014).

5 Oxfam/shelterproject determine seven phases of a transitional settlement strategy's 'operations and planning'. These are the 'preparedness phase'; 'contingency phase'; 'transit phase'; 'emergency phase'; 'care and maintenance phase'; 'durable solutions phase'; and 'exit strategy phase' (Corsellis and Vitale, 2005: 40–47).

6 The Z-score is 'the deviation of an individual's values [BMI/weight-for-height] from the value of a reference population taking into consideration the standard deviation of the reference distribution' (UNHCR, 2007: 297). Individuals are produced a set of biological calculations (standard deviations, distributions) purely defined in relation to the population as a whole.

7 Significantly, Syrians fleeing to Turkey are considered 'guests', not refugees. Their camps are officially 'guest camps' rather than 'refugee camps' (Özden, 2013: 5), though moves have been made since 2011 to give Syrian refugees a clearer legal status (Kirişci, 2014: 14). This irregular status gives the Turkish government a more complete control, restricting the influence of the UNHCR. It has also led to better conditions in some camps than the UNHCR provides (Kirişci, 2014: 15; McClelland, 2014).

3

# Flourishing Hospitality: Global Cities

A much greater proportion of refugees look to the refuge provided by the anonymity and freedom of the city than the humanitarian government of camps. Based on incomplete figures, UNHCR estimates that 56 per cent of refugees in 2013 were residing in urban areas, around 90 per cent in private accommodation (2014b: 37). The Turkish government claims that, during the ongoing crisis, 64 per cent of Syrian refugees in Turkey live in cities, outside of camps (AFAD, 2013: 18). Of course, the distinction between camps and cities is not clear cut; many planned camps are within cities and can become indistinguishable from the surrounding urban environment (Peteet, 2005; Sanyal, 2011, 2014). Current Turkish camps, for example, mostly reside in or very near cities (AFAD, 2013: 16–17). It is also increasingly common to characterise the types of sociality, economy and architectural development of camps as proto-urban or 'becoming-city' (de Montclos and Kagwanja, 2000; Agier, 2011; Herz, 2013; Sanyal, 2014). Nevertheless, the humanitarian hospitality of the camp is often rejected by displaced people in favour of the much less caring, but also *perhaps* less controlling and containing hospitality of the city.

The hospitality afforded by urban spaces has a different character to that of the camp, though it is related through the mechanisms of security which help to structure both. Cities, like camps, have always been produced as spaces of hospitality, but that hospitality is not humanitarian; it works towards different ends via a less totalising ethos. When we think of spaces as spheres of coexistence, where various unrelated trajectories cross or connect, cities are the most open space discussed in this book: 'peculiarly large, intense and heterogenous constellations of trajectories, demanding of complex negotiation' (Massey, 2005: 155). No longer strictly 'bounded', the city is 'relationally constituted, a space where multiple geographies of composition intersect, bringing distant worlds into the centre of urban being and projecting the placed outwards through myriad networks' (Amin, 2012: 64). To the extent that an ontology of the urban is possible, it must be an open one (Amin and Thrift, 2002: 28).

So-called 'global' or 'world' cities are a peculiar case in this regard. On the one hand, they require openness to mobilities and flows of many different kinds. With the rise of globalisation, global cities such as London, New York and Tokyo become crucial nodal points of command and control, channelling flows of goods, peoples and services (Sassen, 2001: 5–6; see also Friedman, 1986; Marcuse and van Kempen, 2000; Taylor, 2004). While there is no agreement on what constitutes a global or world city, its openness to strangers is a necessary if not definitive requirement. Yet, cities do not welcome all strangers equally. A hierarchical welcome operates, not over entry to the global city, but over strangers' movement around it, their sense of ease, belonging and 'at-homeness' within it. Some flows are better than others: some strangers are more welcome; some are not welcome at all; others are silently and invisibly welcomed, allowing for their greater exploitation and control. Yet the hospitality that the global city requires, indeed the hospitality that produces it as a particular type of space, also allows for a number of important counter-conducts and redirections of its hospitality.

This chapter proceeds by outlining the ethos that defines how cities, particularly global cities, welcome migrants. Unlike camps, the urban ethos is one of freedom and flourishing, but nonetheless operates via tactics of security that promote some circulations and restrict others, creating different subjectivities of host and guest in the process. The second section outlines how such a flourishing hospitality has scripted the space of one particular global city, London, by examining key planning and strategy documents produced by the Mayor of London (MoL) and his Office between 2009 and 2014. Finally, I turn my attention to the way strangers and guests in the global city have used it, seized its space, or redirected it to their own ends. Such counter-conducts can take different forms and are not necessarily 'good', but they are enabled by and themselves undergird the openness and ungovernability of the city. Addressing implicit and explicit counter-hospitalities is a way of testing the geographies of (ir)responsibility contained in the global city's empowering and exploitative offer of flourishing hospitality.

## THE URBAN ETHOS, HOST AND GUESTS

This section does not attempt to give another definition of what a city *is*. Many urban theorists and human geographers have contributed to this contested discussion and I have little to add.[1] They are defined by their materiality (the buildings, roads, cars and other infrastructure), the sheer number of people and their interactions, the density of their dwelling together, the heterogeneity of those people's social make up (in terms of culture, race, gender, class, religion, etc.), the everyday interaction between the human and non-human as they flow and stall, the unexpected, surprising and volatile nature of those interactions, and more besides. The 'wonderful thing is that the contemporary city can be understood in all these ways, and yet is not reducible to any one of them' (Amin and Thrift, 2002: 48). It does not give itself up to coherent or totalising conceptualisation because it has no internal coherence:

> The city's boundaries have become far too permeable and stretched, both geographically and socially, for it to be theorized as a whole. The city has no completeness, no centre, no fixed parts. Instead, it is an amalgam of often disjointed processes and social heterogeneity, a place of near and far connections, a concatenation of rhythms; always edging in new directions. (Ibid.: 8)

However, it is this very permeability that is of interest to a study of migration and hospitality. What could be more hospitable than a permeable space with no internal completeness or fixity? What space could welcome with fewer restrictions, whose lack of fixity allows guests to make of that space what they will?

Yet this is also what makes urban space deeply problematic. As noted in the Introduction, hospitality is a structural and affective spatial relation, demanding an inside, an outside, and the rigorous delimitation of boundaries. A home needs impermeable walls, but also doors and windows that open to allow welcome. Yet those doors and windows must also close, allowing the home to be secured. It requires that those able to choose openness or closure, the hosts, feel a sense of belonging, sharing an ethos, a way of being and dwelling that defines their relation to others outside. If the boundaries of the city are so permeable, can it exclude? Can it control its own thresholds if they are so indiscernible? Without centre, fixity or completeness, how can a city produce a sense of belonging, its own ethos? Who is the 'host' of such a space? The nature and permeability of a city's boundaries differ along a spectrum of indiscernability. Some are relatively clearly defined (e.g. Hong Kong) while most others leak out into the locality and wider world through networked connections to the global-urban (Magnusson, 2011). This is why I focus on one example below, exploring how London has produced and scripted itself and its boundaries as hospitable. However, it is important to say something more general about the ethos of cities, a specifically urban way of life, and how this works to construct the identities of hosts, guests and those falling in between.

## THE URBAN ETHOS

Jonathan Darling (2013: 1785) observes that what he calls 'moral urbanism – the discursive and affective construction of particular cities as imbued with moral characteristics' is distinct for different cities. It is certainly possible to see different cities embodying (or at least marketing themselves as embodying) a certain set of values and principles. But this is not the same as an urban ethos, a way of life common to cities. Many of the aspects we may associate with the urban, some of which were mentioned above, mean the city has often been conceived as a space of opportunity, unexpected encounter and, ultimately, *freedom*:

> [T]he city is widely perceived as a locus of freedom: a place where the enclosures of family and tribe and tradition can be escaped and where new modes of life – perhaps ones that refigure the family, tribe, and tradition in new ways – can be created. In short, urbanism as a way of life is a form of human freedom: for many people, the ultimate form of human freedom. (Magnusson, 2011: 22–23)

The city's ethos of freedom is one that welcomes all forms of life, whether consciously or not. It is hospitable to difference, to a plurality of ways of being. Hence, cities are often seen as 'spaces in which difference proliferates and mixes ... in which agonistic interplay is a constitutive feature, the city is defined by plurality (even if its citizens regularly try to disavow it)' (Coward, 2012: 471). As such, they are spaces where all forms of life can flourish, where everyone can be 'at home' and find some sense of belonging. Those individuals or groups who feel marginalised and excluded from rural and traditional communities can find a place to 'be themselves' in an urban context. A classic example could be the Castro district of San Francisco, an inner-city area which is treated as 'home' by a transnational gay community, even if they do not reside there permanently (Duyvendak, 2011: 62–82). But urban community is neither necessary nor necessarily easy to access. A key aspect of the city's liberating ethos can be felt as, or characterised by, solitude, anonymity and loneliness in the density of the crowd. Hence this ethos includes 'a critical negative freedom, the freedom that follows in part from the indifference of others – the right to be left alone'; cities' ethos and hospitality are as much about an 'ethics of indifference' (Tonkiss, 2005: 28) as they are a welcoming embrace of plurality.

This is, admittedly, a highly romanticised and Western view of the urban. Cities, like refugee camps, can be sites of great violence, exclusion and hostility, whether experienced through everyday racism, sexism, class warfare and homophobia, or the concentration of the poor in slums (Davis, 2006; Koonings and Krujit, 2007). Nor are these unfreedoms the preserve of cities of the global South, as spectacular riots by the urban 'underclass' across Europe and North America attest (Murray, 2006; see Malmberg et al., 2013; Sokhi-Bulley, 2015). However, my point in identifying this ethos of freedom is two-fold. On the one hand, its veracity is in a sense unimportant; what matters is that this is a recognised regime of truth seized upon by cities, their promoters and planners. Thus, when former London Mayor Ken Livingstone responded to the London bombers in 2005, what he identified as key to London's ethos was the freedom and the flourishing it allows:

> [People] choose to come to London, as so many have come before because they come to be free, they come to live the life they choose, they come to be able to be themselves. They flee you because you tell them how they should live. They don't want that and nothing you do, however many of us you kill, will stop that flight to our city where freedom is strong and where people can live in harmony with one another. (Closs Stephens, 2007: 168–169)

Such a vision of London's hospitality blithely effaces the past and present colonial violence upon which London's multicultural freedoms and welfare are built (Closs Stephens, 2007: 170; Massey, 2007: 175). But its ethos remains intact and becomes a truth upon which cities plan and articulate their relation to others.

On the other hand, this liberating ethos is deeply implicated in the way the global city is governed through tactics of security. This does not mean that violence and exclusion are eradicated; far from it. Rather, these security mechanisms by which the other is both welcomed and managed aim to govern *though* freedom, to enable circulations within 'ever-wider circuits', to 'let things happen' at a local level in order to prevent a general problem arising to endanger the whole population (Foucault, 2007: 44–47). A certain optimal level of safety is thus the aim, within a 'bandwidth of the acceptable' (ibid.: 6). The urban population is therefore in large part left to self-govern, to self-regulate, managing each other's behaviour in their own 'security regime' (Magnusson, 2011: 32). Authorities of various kinds proliferate – from local government to community and charity organisations, churches and mosques – creating the means by which the urban population watches over its own behaviour. Governing through freedom multiplies 'the points at which the citizen has to play his or her part in the games that govern them' (Osborne and Rose, 1999: 752). The genius of urban self-government is that it allows for the extemporaneity of cities, the freedom to encounter the unexpected, whilst bringing it under a degree of control. Its 'governability' therefore arises 'out of its spontaneous, ungoverned features' (ibid.: 758).

This ethos of free flourishing generates a space of differential welcome, containing a sense of belonging and non-belonging. 'Urbanism may be a form of freedom, but it is nonetheless totalitarian. It tends to extinguish forms of life that are incompatible with it' (Magnusson, 2011: 23). Thus, the openness of London's cosmopolitan hospitality trumpeted by its former Mayor contains a closure to those who do not endorse the values of diversity and multiculturalism. Hubbard and Wilkinson (2015: 606) illustrate how the London Olympics' insistence on being open to lesbian, gay, bisexual and transgender (LGBT) populations meant that it 'failed to extend a welcome' to those who did not uphold 'particular "civilized" norms'. The 'intolerant homophobe' does not belong, and is thus not free to 'live the life they choose' in the global city. This, of course, can operate as an excuse for related intolerances such as Islamophobia (ibid.: 606). An even firmer rejection of non-belonging came in 2009, when recent Mayor Boris Johnson responded to claims from neo-fascist politician Nick Griffin that London had been 'ethnically cleansed':

> Nick Griffin is right to say London is not his city. London is a welcoming, tolerant, cosmopolitan capital which thrives on its diversity. The secret of its long-term success is its ability to attract the best from wherever they are and allow them to be themselves – unleashing their imagination, creativity and enterprise. The BNP [British National Party] has no place here and I again urge Londoners to reject their narrow, extremist and offensive views at every opportunity. (Mulholland, 2009).

Here Johnson privileges hospitality as the causal factor behind London's success as a global city. He does not exclude the BNP through sovereign mastery of London; rather, he 'urges' Londoners to self-manage the space, *and themselves*. 'Londoners', those who demonstrate their belonging by practising an ethos of tolerance and freedom, do so precisely through the intolerance and rejection of that which does not conform to this ethos, that which 'has no place here'.

## HOSTS AND (G)HOSTS

It is significant that Mayor Johnson does not place himself as the sovereign host who finally decides on what is to be welcomed and excluded. When Magnusson urges us to 'see like a city', rather than like a state, one of the key implications is that 'sovereignty is infinitely deferred' (2011: 7). In its place we see processes of governmentality which operate via tactics of security described in part above as constitutive of an urban ethos. Rather than a host that decides which strangers are to be welcomed as guests and which are to be indefinitely excluded, cities' self-organisation involves a 'multiplicity of authorities in different registers and at different scales', where practices of control are temporary, local and open to unpredictable transformations. As in refugee camps, the host is not a state, agency or single body; it is an assemblage of this urban multiplicity. The global city forms out of this ensemble where hosting practices constitute a never fully-formed, always changing 'host' that rejects, welcomes and governs migrants.

We can see this temporary assembling in the account 'London' gives of itself. Under the title 'Who Runs London', the Greater London Authority (GLA) website noted in 2010 that 'We have our own unique way of running London. It involves a number of key players with different roles and responsibilities and a shared commitment to making London the best city in the world' (GLA, 2010).[2] This 'we' that runs London, that is ultimately its host, is present only in being ungraspable. While much of the material I examine below comes from the Office of the Mayor of London, the Mayor's role is only one of 'strategic development', setting the 'vision for how to make London an even greater city', encouraging and backing action to 'realise that vision' (ibid.). S/he cannot operate without the extensive support offered by the GLA and the London Development Authority (LDA); the actual delivery of policy is largely handled by the 32 London boroughs, the City Corporation of London and central government. Emphasising the transformations spoken of by Magnusson, the

LDA has since been closed and many of its functions have been brought under the GLA. A map to account for the new arrangement and answer the question of 'Who Runs London?' has been produced by the London Communications Agency (2011).[3] But this 'is a living document, updated regularly' with 'changes to key public sector agencies at local, regional and national levels in London'. The host is never complete, never fully present, always assembling and disassembling; it is spectral in its non-presence, its non-completeness.

But the governmental structure of the assembling host proliferates further. The Mayor also works 'closely and collaboratively with a wide range of public and private institutions' in the formation and delivery of policy (LDA and MoL, 2010: 36). Every framework and strategy which produces London as a welcoming space and outlined in London's planning documents, involves working with QUANGOs, non-ministerial government departments or private bodies. For instance, the Ambient Noise Strategy, which aims to make London a welcoming 'soundscape', requires collaboration with central government, the Highways Agency and Heathrow Airport (MoL, 2014a: 260). Indeed, despite the continued emphasis on the centrality of culture and the creative industries in making London an exciting, enticing space for guests – they are 'the key reason why people visit' and 'move to' London (MoL, 2008: 17) – 'no single agency for culture exists in London, and nor is the GLA a major direct funder of culture.' There is no host that fixes or organises the culture or ethos of London, defining its reception of the world, just an ever-changing ensemble.

The spectrality of hosts does not prevent global cities from being hyperactive in attempting to welcome migrants from all over the world. Indeed, it is a hallmark of global cities that they compete with each other to offer a more hospitable environment for business and leisure. One obvious example of this competitive hospitality is over the staging of 'mega-events' (Roche, 2000), such as trade fairs, expos and sporting events like the Olympics. Since 2004, the ability to 'welcome the world' by hosting the Olympics has been key to all aspects of London's present and future hospitality (see Brassett, 2008; Bulley and Lisle, 2012; Hubbard and Wilkinson, 2015). In a more mundane manner, cities compete on an everyday basis to attract tourists in the lucrative 'city-breaks' market (Knox, 2011). But they also seek to out-host each other by appealing to the cosmopolitan workforce. Richard Florida famously claimed that this 'creative class' is a self-selecting group 'who adds economic value through their creativity' – scientists, artists, IT specialists, software programmers and architects, amongst others (2002: 68–69). His controversial thesis (see Peck, 2005) argues that companies must locate in cities that attract these 'no-collar' workers. The most successful, like San Francisco, New York, London and Sydney, are 'places that are open to immigrants, artists, gays, and racial integration' (Florida, 2005: 7). Cities that thrive in the global economy are those that promise the greatest flourishing; the best hosts are those that are tolerant and open, most clearly embodying the urban ethos of freedom and its relation to migrants.

While there need not be a singular 'host' to ensure and govern this welcome, the idea of cities as spaces of hospitality begs the question of who performs the everyday acts of hosting. Who carries out the mundane tasks of servicing, feeding, sheltering, cleaning and securing those temporary and semi-permanent creative guests which global cities compete to attract? This issue of embodied hosting refers to what Saskia Sassen calls a 'structural process': global cities are a 'key site for the incorporation of large numbers of immigrants in activities that service the strategic sectors' (2001: 322). As global cities compete to welcome guests who sustain their global competitiveness, they also require a constant flow of less-valued workers to provide traditional hosting services. Private households are developing with no traditional 'wife' figure, leading to a 'return of the so-called "serving classes" in all the global cities around the world, made up largely of immigrant men and women' (ibid.: 322). 'Talented' and 'creative' global guests are welcomed into cities not by the host, but often by other guests whose 'lesser' skills are vitally important to maintaining urban cosmopolitan openness. Despite their hosting activities in low-paid service jobs such as catering and cleaning (literally, the 'hospitality' industry), they are ostensibly guests of the global city.

Such guest-hosts, or '(g)hosts' (Bulley and Lisle, 2012; Bulley, 2013), embody the spectrality, the absent presence of the host. While Sassen (2001: 305–325) stresses the role that (g)hosts play in New York, London is also 'seriously dependent for its normal functioning on labour from elsewhere' (Massey, 2007: 175). Simply to sustain its own health and deliver its babies, London requires nurses from countries in Asia and Africa which paid to train them. Sri Lanka and Ghana are effectively 'subsidising the reproduction of London' (ibid.: 175). But such parasitic reliance on foreign nurses is symptomatic of a much wider dependence. Official data from 2001 suggests that up to 46 per cent of those doing less 'skilled' hosting work in the city (e.g. domestic labour, caretaking, refuse collection and cleaning) were not born in the UK. The majority are from poorer parts of the globe (Spence, 2005). The 'hospitality industry' is most dependent on such immigrant labour, with rates of well over 50 per cent (Matthews and Ruhs, 2007: 36). In 2004–5, it is thought that 76 per cent of chefs and cooks in London, and 69 per cent of cleaners, were foreign born (Wills et al., 2009: 263). As the next section examines the official production and government of London as a space of hospitality, it is important to search for these (g)hosts and their place within London as an open and tolerant home.

## PRODUCING LONDON BY WELCOMING THE WORLD

In this section, I am interested in drawing out the way London has 'officially' produced itself as a space of hospitality through the plans and strategies that it publishes and amends throughout a mayor's tenure. The generation and alteration of some of these, such as *The London Plan* (or Spatial Development Strategy), are a legal obligation under the legislation that established the GLA (MoL, 2014a: 5). Early drafts of the current Mayor's Plan – a product of the Mayor's Office, the GLA and the now defunct LDA – were published for public consultation in 2008 before being finalised in 2011 and altered several times up to the most recent version I will look at, published in January 2014. Others, such as the *Economic Development Strategy* (LDA and MoL, 2010), the *Tourism Action Plan* (LDA and MoL, 2009) and *Cultural Strategy* (MoL, 2014b and 2008), are not altered as regularly. The latest draft of the *London Plan* includes changes as a result of Mayor Johnson's 'vision' document (MoL, 2013). All this planning and strategising may seem a little excessive, but the aim is clear: 'we want London to be the best big city to live in – not just because that is the best thing for Londoners, but because we are competing with rival cities around the world' (MoL, 2013: 48). Who this 'we' is remains unclear, but these documents are a key intertext for investigating the construction of London as a particular kind of open, cosmopolitan and exploitative home.

## A (NEARLY) UNIVERSAL WELCOME

London is scripted as a universally welcoming home. It is promoted as 'an open and welcoming city' because 'openness to people and ideas from around the world' is 'an economic asset' and 'a key element in generating innovation and growth' (LDA and MoL, 2010: 29). Proudly declared to be the most visited city in the world (MoL, 2013: 66), its hospitality is central to shaping London as a 'home' along with all its spatial, cultural, transport and tourist plans. If we look at the 'Key Diagram' in Figure 3.1, which brings together the 'main components' of its spatial strategy, we see a visual representation of this openness, with five separate entry and exit points crossing the boundaries of 'London' and designated as 'regional coordination corridors'.

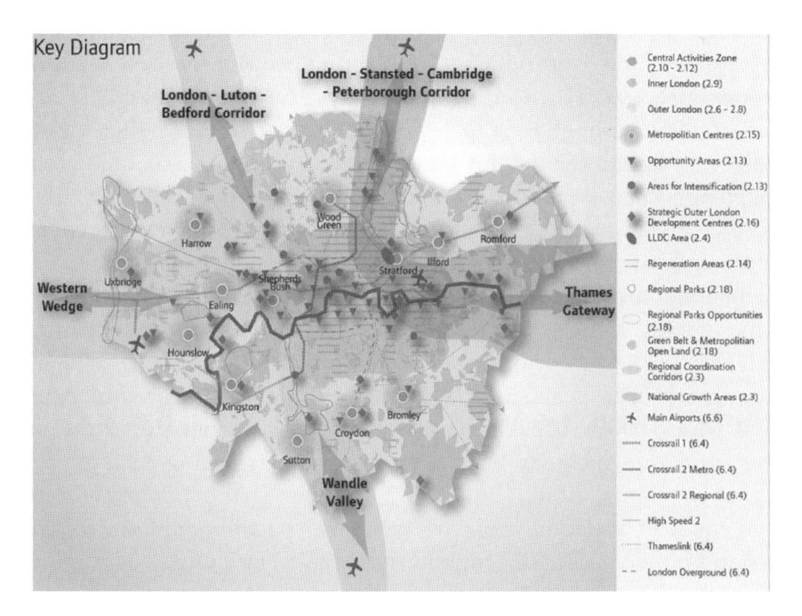


**Figure 3.1** Corridors to the London home and its hospitality (MoL, 2014a: 78)

However, while these corridors leading in and out of the urban home could appear as doors, they also show the blurring of London's boundaries. After all, three lead outwards towards 'London's' own airports, themselves entry and exit points to London. Only two of London's airports (City and Heathrow) are inside the boundaries of 'London'; London Luton and London Gatwick are in Bedfordshire and Sussex respectively (the 'home counties'), while London Stansted is nearly 40 miles away in Essex. Thus, the corridors are both inside and outside the home. London thus becomes blurred, it 'seep[s] out at the edges' (Rose, 2000b: 95), its windows and doors appearing a long way from its shifting walls.

Not only is London portrayed as welcoming and incredibly diverse, it is considered to always have been so, and must become *more* so. Hospitality is its history and its destiny. These documents offer a narrative of London's past 2000 years as a space which has been a 'home for people in all walks of life and from all parts of the world' (MoL, 2014a: 11). Its peculiar economic success is a 'continuation of its remarkable history' of 'being open to the world' (LDA and MoL, 2010: 25), 'for centuries ... the crossroads of the global economy' (MoL, 2013: 66). Even its built heritage exhibits the 'story of the city' as multicultural, yet cohesive (MoL, 2014a: 247). And London intends to build on this, seeking to be 'even more open' (MoL, 2013: 64).

This openness is also evident in the 'guest' that London seeks to attract. While refugee camps generate a guest population to be fixed and measured, London is far less publicly concerned about predetermining its guests. Thus, while the Mayor wants immigration policy which will 'attract the brightest and the best to London but keep out those who have no intention of making a contribution' (MoL, 2013: 51), who 'those' non-contributors are is left as a blank. They are not dangerous, merely unproductive. The 'brightest and the best' is fleshed out by 'London First' (2015), an influential lobby group of businesses. They target two types of guest, campaigning for 'immigration and border policies that enable global talent to work and study in London, encourages high-spending tourists to visit and ensure that all visitors are made to feel welcome'. Though 'all visitors' are welcome, 'global talent' and 'high-spending tourists' are actively encouraged. Meanwhile, (g)hosts remains invisible and unmentioned.

Though London does not seek to construct a guest population in so meticulous a manner as refugee camps, its welcome is still carefully planned and managed. One of the top priorities of the *Tourism Action Plan* is to 'deliver and promote a world class sense of "Welcome" throughout the visitor experience' (LDA and MoL, 2009: 20). Hospitality is promised 'from pre-arrival to post-departure' and this includes 'inspiring all customer facing staff to raise the standard of welcome' (ibid.: 26). More concretely, the London Ambassador Scheme includes a '"welcome" role' for personnel 'patrolling the streets, transport gateways and public spaces of London' (ibid.: 26). The chances of coming across such a 'London ambassador' must be rather small, with only 300 having been trained by 2009 (LDA and MoL, 2009: 18). A guest's embodied reception is more likely to come from the hospitality industry, through its acceptable face of taxi drivers, hotel managers, maids, cleaners, chefs, waiters, security staff and souvenir sellers, or its less acceptable face of sex workers, traffickers and drug dealers. Either way, the face of welcome is more likely than not a foreign-born face, a guest themselves. Yet such guest-hosts are not mentioned in London's planning documents. They are not Johnson's unproductive, unwelcome strangers; neither are they necessarily 'Londoners'. They are certainly not the prized 'global talent'.

The irony is that (g)hosts also make London's hospitality possible in a second sense: they help constitute the diversity that the city parades to the world. A constant refrain is that London's 'diversity is one of its greatest strengths ... more languages and cultures are represented here than in any other major city' (MoL, 2014a: 81–82). Its 'diverse culture' is what attracts so many talented people from all over the world (LDA and MoL, 2010: 18), and it 'draws strength from the immense variety of its people' (ibid.: 28–29). (G)hosts, those who formed this diversity over many years and, still arriving, continue to constitute it, are finally revealed here in London's hospitality script. But they appear only as a backdrop. They emerge as a contextual attraction, something the 'brightest and the best' can enjoy sampling or experiencing, like a stroll around the Tate Modern. The only other way London's dependence on invisible (g)hosts is revealed is through an understated attempt to make the 'best' of them *less* invisible: the Mayor advocates 'an earned amnesty for irregular migrants who have been law abiding and working in London for a number of years'. This is supported by an LSE study which estimates that an amnesty could add £3 billion to the UK's GDP (LDA and MoL, 2010: 49).

So London's hospitality is dependent upon legal, illegal and semi-legal migrants in two ways: to provide the diversity that enables its scripting as genuinely open and welcoming; and to carry out the material work of hosting, despite often being guests. This productive confusion of hosts and guests, itself a seemingly malign form of hospitality, helps reproduce the most efficient governmentality of hospitality. The use of (g)hosts makes London both a cheaper host and a more welcoming space, as its diversity means it is a place everyone 'feels at home'. It is not in the interests of the global city to prevent migration; its governmentality is not targeted at restricting movement but 'the opening up and release of spaces, to enable circulation and passage' (Elden, 2007: 30). The urban conduct of conduct therefore works with the ethos of global cities, *facilitating* rather than impeding circulation, though only along acceptable channels. The 'earned amnesty' for good migrant (g)hosts would work in this vein, enabling more efficient circulation of (g)hosts who could also pay taxes.

## ENCOURAGING ACTIVITY AND MOBILITY

An important way in which the urban home is imagined and governed is the separation of 'private' space and 'shared' space. This division is mapped from the Mayor's foreword to the 2009 consultation draft of the *London Plan*, where

Johnson suggests that the focus of planning strategies is on shared space as the foundation of the urban – its freedom and opportunity. 'This shared space is a vast and complex environment in which millions of perfect strangers must move, meet and negotiate ... The genius of a big city is in the way it organises that shared space, for the benefit of visitors and inhabitants alike' (MoL, 2009: 5). It is the intricacies and multifaceted nature of such space, as well as the attempts to 'organise' it, that works on host and guest to render them both governable and self-governing. 'The city itself transforms those who reside in it, so that even the most culturally bound are impelled to negotiate the complexity of the city and the diversity of its populations' (Jabri, 2010: 55). Not only does this complexity arise from the diversity of people, but also what is 'between them', the dense and interconnected materiality of the city that divides and connects them (Coward, 2012).

Shared space is therefore where London's hospitality is managed and controlled. In Chapter 2 of the spatial plan, entitled 'London's Places', the city is mapped and re-mapped in various different ways to show further subdivisions and facets of its ordered welcome. We see London's sub-regions (MoL, 2014a: 46, Map 2.1), its development sites (ibid.: 61, Map 2.4), its regeneration areas (ibid.: 63, Map 2.5), its 'strategic industrial locations' (ibid.: 73, Map 2.7) and its network of 'strategic open spaces' (ibid.: 76, Map 2.8), before reaching the 'Key Diagram' in Figure 3.1. While all these mappings of shared space are important, and much could be said of each, I concentrate on the firmest sub-division of space, which is referred to through the Plan and included in the 'Key Diagram': the division between Outer London, Inner London and the Central Activities Zone (CAZ) shown in Figure 3.2.

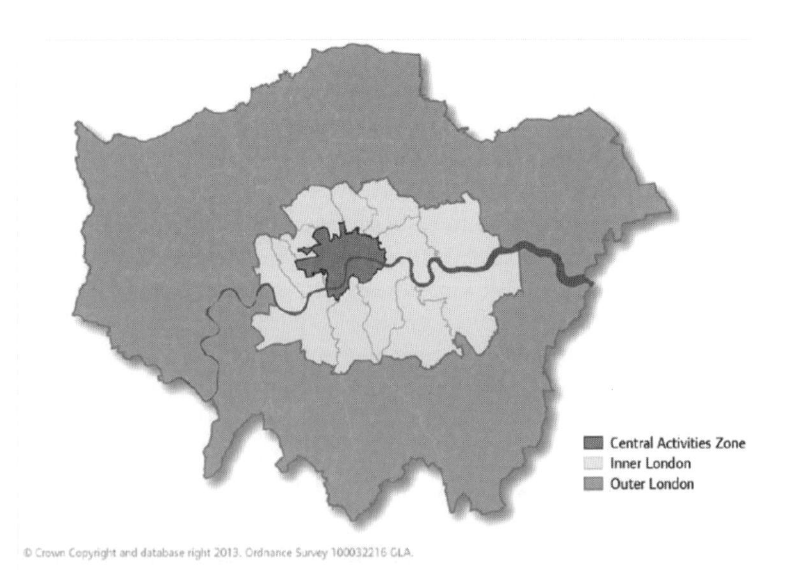


**Figure 3.2** London's zones of hospitality (MoL, 2014a: 47, Map 2.2)

This particular mapping graphically underlines (pictured in bold, darker shade) the dominance of central London in its vision of itself as a home. The CAZ is London's 'geographic, economic and administrative core' (2014a: 56). It contains the 'largest concentration of London's financial and globally oriented business services', but also 'embraces much of what is recognized around the world as iconic London'. It is the centre of London's hospitality because it attracts *both* the key guest demographics identified by the spectral host ('high spending tourists' and 'global talent'). Thus the CAZ has become 'the world's leading visitor destination' (ibid.: 55). Meanwhile, it is 'home' to only 284,000 people, a mere 3.5 per cent of the 8 million-plus Londoners (ibid.: 55). In contrast, 60 per cent of Londoners live in Outer London (ibid.: 46). The CAZ is thus the space of activities and entertainments, a zone primarily for guests rather than hosts.

Inner London, meanwhile, contains 'probably the largest concentration of deprived communities' and some of the 'most challenging environments', including a hugely varied ethnic population, high housing densities in high-rise estates, outdated social infrastructure and limited access to open space (ibid.: 52–54). While by their very nature they are invisible, or semi-visible, this is the most likely 'home' of the (g)hosts that do the embodied work of hospitality – deprived parts of Inner and Outer London where housing is cheapest and existing diaspora communities can help with accommodation and finding work (Wills, 2012: 118). The importance of the space for the privileged guest as opposed to the deprived hosts and (g)hosts is emphasised by the fact that the CAZ is accorded four and a half pages of policy and analysis, whilst Inner London receives one and a half. Inner London boroughs operate by the usual planning procedures, but those 'with all or part of their area falling within the CAZ' must develop 'more detailed policies and proposals' which take account of the CAZ's priorities and functions (MoL, 2014a: 56).

The deprived inner city is not central, 'iconic London', and contains little that the planners would like either demographic of the desired touristic or talented guests to see. It forms the boundary of the acceptable home; while few hosts actually reside in 'iconic London', it is devoted to hospitality and hosting practices. The CAZ is effectively the 'good room' for receiving guests, where the host does little but clean and polish; inner London is the grubby utility room and understair cupboard, home to the machinery of hospitality; outer London is the functional living space of the home, the site of the mundane and everyday. An inner division is thereby created which both includes and excludes, governing the movement and choices of both hosts and guests by keeping key attractions within the CAZ. However, this planning and governing of mobility and activities is designed to work so that guests do not know or feel their mobility and activity being managed. They simply go where the attractions are. These mappings are not really creating new divisions, but reflecting and justifying a material reality. As Mayor Johnson notes, 'the main reason talented people choose to work in London is the "vibe"; and though there is not much that politicians can do to create "vibe", they can create the conditions for these good vibrations' (MoL, 2013: 64). But by planning and strategizing in this manner the host helps to cement and naturalise the conditions in one area *in particular*. The claim that the CAZ has a special 'vibe', or a 'unique character and feel' (MoL, 2014a: 56), thus helps constitute the material reality, entrenches its divisions, contains activity and operates to limit spontaneity and spatial creativity.

This constituting of vibe, with its subtle government of activity, is also apparent in the differential classification of 'town centres' which London offers its guest and host population. These are divided into four types, mapped in Figure 3.3: international centres, metropolitan centres, major centres and district centres. International centres are 'globally renowned retail destinations' with excellent public transport connections (MoL, 2014a: 317). There are only two of these and they are, unsurprisingly, in the CAZ, the boundary of which divides the 'cosmopolitan' space of urban hospitality from the 'metropolitan' town centres. This second category merely has a wide catchment area of several boroughs (ibid.: 317), and the majority of these are in Outer London. Inner/Outer London is not marked on this map, denoting that the CAZ/rest of London is the most important border.[4] Similarly, elsewhere the mapping of London's 'night time economy of strategic importance' (bars, restaurants, nightclubs and performing arts venues) designates only two areas of 'international importance', both in the CAZ (see MoL, 2014a: 142, Map 4.3).

The determination of 'Strategic Cultural Areas', identified in order to 'protect and enhance' places with significant clusters of cultural institutions,

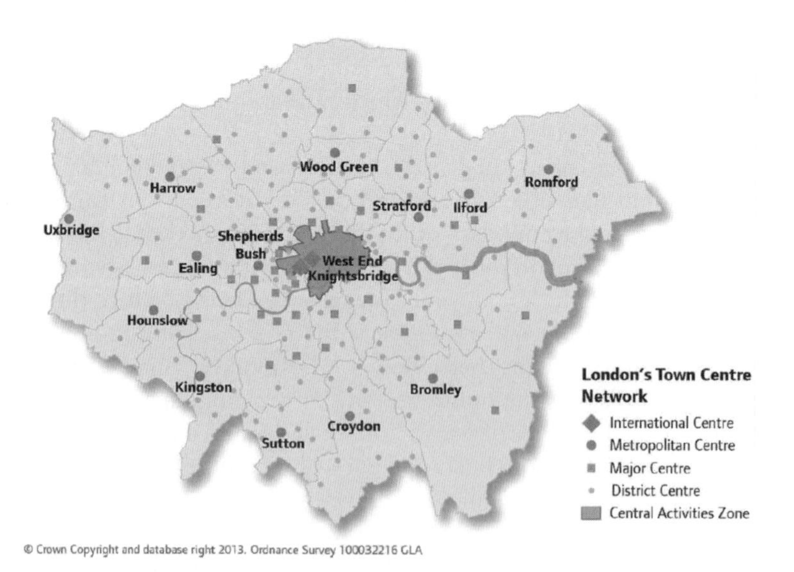


**Figure 3.3** London's centres of activity and 'vibe' (MoL, 2014a: 66, Map 2.6)

reiterates the divide. Of nine such areas, four are concentrated in the CAZ (West End, South Bank, Barbican and South Kensington Museum Complex) with two warranting trips outside for specifically sporting reasons (Wembley and the Olympic Park) (see MoL, 2014a: 138, Map 4.2). According to London's spatial plan, tourists and talented guests will not be encouraged to venture outside the CAZ; if they do so then accommodation planning should be 'focussed in town centres and opportunity and intensification areas, where there is good public transport access to central London' (ibid.: 136). This will allow guests to return quickly to where their activity really belongs. The welcome offered by London then is not about 'spontaneous movement', but the transportation of guests through a 'series of packaged zones of enjoyment, managed by an alliance of urban planners, entrepreneurs, local politicians and quasi-governmental "regeneration" agencies' (Rose, 2000b: 107).

Transport is thus one of the key ways in which the guest's mobility and activity is encouraged and constrained. Many plans have been put in place for upgrading London's transport, ensuring easy mobility into, out of and around London. This includes automation of 75 per cent of London's Tube network by 2020 and a vast increase in capacity on key lines, development of important train, tram and cycling networks (MoL, 2013: 16–17). The most ambitious of these projects is Crossrail, described on its website as 'Europe's largest construction project' (Crossrail, 2014). Due to open in 2018 for the 'central section', which is underground, with five new stations in the CAZ itself (MoL, 2014a: 66), stations will gradually extend out to Reading and Heathrow in the West and Shenfield in the East. Twenty-four trains are planned to run on Crossrail every hour in each direction at peak times, 'transform[ing] public transport in London ... and bringing an extra 1.5 million people to within 45 minutes of central London' (ibid.). Plans are already in place for Crossrail 2, which would connect the South East with the North West home counties, all through the CAZ (see MoL, 2013: 78–79). These Crossrail plans are not, therefore aiming to connect Reading and Epsom with Shenfield and Cheshunt, but to allow speedier access for people in all these places to the CAZ, where everyone effectively becomes a temporary guest.

Such grand plans are important, but London's space is managed in many more mundane ways to make it more welcoming while promoting mobility and activity. The *London Plan* calls on *all* 'places' and 'spaces' to be well designed; everyone 'should be able to safely and easily move around their neighbourhood through high quality spaces, while having good access to the wider city' (MoL, 2014a: 236). The outcome should be places 'where people want to live and feel they belong, which are accessible and welcoming to everyone' (ibid.: 238). These are fairly vague suggestions, but the importance of welcoming, mobility-enabling spaces is pursued through specific strategies such as the Air Quality Strategy (reducing public exposure to pollutants – ibid.: 257) and the Ambient Noise Strategy (mentioned above – ibid.: 258–259). There is a London View Management

Framework to help restrict building and planning in order to preserve 'strategic' views of the city and its major landmarks (ibid.: 249). And through the London Tree and Woodland Framework, the Mayor will even advise on strategies for trees that, we are reassuringly informed, will always follow the principle of 'right place, right tree' (ibid.: 270).

Interestingly, traditional security concerns are largely absent from these documents. Precautions against terrorism are almost unmentioned. The host's role seems to be one of merely 'promoting', 'enhancing' or 'facilitating' greater security of train stations (MoL, 2014a: 51), leisure spaces (ibid.: 65), work spaces (ibid.: 130) and London's energy and water supply (ibid.: 157, 178). We are told that the 'Mayor will work with relevant stakeholders and others to ensure and maintain a safe and secure environment in London that is resilient against emergencies including fire, flood, weather, terrorism and related hazards' (ibid.: 255). The matter then seems to be passed to the London Resilience Partnership (ibid.: 256). It is as though the public face of the city turns a blind eye to exceptional insecurities, allowing the space to be secured through visible and invisible restrictions on mobility, the 'rings of steel, rings of concrete and rings of confidence' (Coaffee, 2004: 201–211). There is, however, a brief focus in the Plan on 'designing out crime' (see MoL, 2014a: 239, Policy 7.3), which hints at the ways in which the urban home is secured. Encapsulating the idea of governmentality operating through security mechanisms, 'routes and spaces should be legible and well maintained, providing for convenient movement without compromising security', aiming to 'reduce the opportunities for criminal behaviour … without being overbearing or intimidating' (ibid.: 239). Buildings should include 'appropriately designed security features' (ibid.: 239), presumably meaning those such as bollards, reinforced concrete planters and strategically placed benches (Coaffee, 2009: 286–294) which the guest will not notice. These will guide their movement, encouraging the good and constraining the bad, without anyone realising their behaviour is being steered. What we are left with is a governmentality which is excessively open and welcoming, but which manages and secures the urban home through tactics that enable and constrain activity without host or guest awareness. Spontaneity, freedom and flourishing is maintained, but moulded to ensure efficient, safe, *productive* circulation around a flourishing CAZ which attracts guests and conceals (g)hosts.

## REDIRECTING URBAN HOSPITALITY

The flourishing hospitality that creates global cities such as London as specific type of home is carefully scripted. Nonetheless, the global city is not the biopolitically managed space of the refugee camp, and is certainly not a disciplinary panopticon. Its governmental management enables a form of freedom, and even those subjects and movements it seeks to control will always escape its grasp.

Its lack of totality and the spectrality of its host makes the city 'a series of partial orders, localized totalities, with their ability to gaze in some directions rather than others' (Amin and Thrift, 2002: 92). And this (dis)ability is hugely productive as it enables the host to construct, constrain and then turn a blind eye to the (g)hosts who maintain its home as enticing, efficient, safe and clean. The global city's 'blind spots' are necessary. However, they also allow possibilities for the evasion and redirection of its hosting and guesting practices.

Such resistance can take a huge variety of forms, from the attempt to seize the 'shared' space meant for guests and 'Occupy' it in a more open, hospitable manner (Bulley, 2016), to the apparently 'mindless' rioting that recreates urban space through lawless spontaneity (Sokhi-Bulley, 2015). It can involve less spectacular seizing of space through migrant workers organising locally on their own behalf (Wills, 2012). I want to focus on two counter-conducts in particular, both of which point to the limits and possibilities of the city as a space of hospitality: the Sanctuary Cities social movement that seeks to make blindness towards (g)hosts a policy of host populations and municipal government; and the practice of transiency amongst some (g)hosts that uses the freedom of the city and redirects it to the (g)host's own particular ends. I begin with the latter.

## GUESTING COUNTER-CONDUCTS: TRANSIENCY AND 'PRECARITY'

The privileged talented and tourist guests of the global city perhaps have no reason to resist the way their conduct is managed inside the home. Why would they want to stray from the attractions and easy circulation of the CAZ, or the delights of Manhattan? They are *able* to visit deprived Inner or suburban Outer London, though there is little to 'see' or 'do' and they will not be made as welcome (with fewer amenities, less information and poorer transport links). Their non-belonging becomes more apparent the further they move from the CAZ. In contrast, the lives of the (g)hosts that do not 'belong' in London and yet (re)produce its welcome are far more constrained: those who are silently welcomed to clean public transportation, offices, hotel rooms and homes; those that secure these spaces, or sell their bodies in one way or another; those that deliver the babies and care for the parents and children of the global cosmopolitan class, cook their meals and clear away afterwards. The way (g)hosts are welcomed and their behaviour carefully managed fashions them into 'precarious workers' (Anderson, 2010), part of a new global sub-class, 'the precariat' (Standing, 2014). This is the place of more urgent guesting counter-conducts.

To seek these out, we must ask what structural factors constrain how such (g)hosts experience the global city, producing them as precarious guests. What tactics of government must they counter in their everyday lives? Here, the spectral host acts through several levels of government. At the national level the deregulation of labour markets in the 1990s and 2000s, the extension of 'guest worker' schemes and the freeing up of subcontracting has combined with increased global mobility to open up flexible, often zero-hour contract jobs to migrants, especially in sectors such as cleaning, construction, food processing and hospitality (Wills et al., 2010: 37). This has been supplemented, particularly in the hospitality industry, through employer practices of 'sub-contracting by stealth' (Evans et al., 2007), with hotels and institutions leaving vacancies unfilled and using lower paid, more easily disposable agency staff as required. This contributes to the insecurity of the migrants: low pay and the flexibility of contracts restricts their choices and ability to plan for the future (Anderson, 2010: 303–304; Burgess et al., 2013).

Subsequently, the points-based immigration system instituted and recently tightened by UK governments has forced legal migrants into certain industries, such as those mentioned above (Wills et al., 2010: 39). If low-point migrants want to move, they must do so silently and invisibly. Meanwhile, the restriction of access to benefits and entitlements further boosts the workforce as asylum seekers and those overstaying their visas must work illegally in the shadow economy to make ends meet. Here they are at their most vulnerable: without the proper papers and credentials, (g)hosts are vulnerable to unscrupulous employers, unsociable hours, hazardous conditions, little or no pay, physical and sexual abuse (Wilkinson, 2012: 16). They cannot complain as such 'hyper-precarious' workers are under constant threat of deportation (Lewis et al., 2014).

In such a situation, one hope is the lifting of the deportation threat through regularisation. Thus, as mentioned, London has lobbied for 'an earned amnesty' for irregular migrants with a good record (LDA and MoL, 2010: 49), a call supported by researchers of precarity to bring the undocumented 'out of the shadows' (Wilkinson, 2012: 19). Yet, even with regularisation, true 'belonging' is not necessarily the result: often migrants' unvalued skills, lack of English and ethnicity make them unattractive to employers – 'When irregular, they were trapped in low-paid employment; and after regularisation, they were trapped in unemployment' (Wills et al., 2010: 56). In 'good' jobs in the hospitality industry, employer prejudices and preconceptions restrict the possibilities for all migrants, segmenting their workforce into roles according to gender, race and national identities (McDowell et al., 2007; Wills et al., 2010).

This is a highly constrained hospitality which exists alongside and enables the freeing, flourishing hospitality offered by the global city. However, those analyses that implicitly construct precarity as containing no choice or capacity for resistance (e.g. Anderson, 2010) are also problematic. In fact, as the research with migrant workers conducted by Jane Wills and her colleagues demonstrates, most have *chosen* to seek out the city's hospitality, and have a number of more or less successful tactics for coping with their exploitation (2010: 121–137). Precarious (g)hosts find a variety of ways to *make use of* the openness, freedom and anonymity afforded by the global city. Though well aware of their exclusion, self-identifying as 'ghost workers' that people don't notice (ibid., 2010: 57, 88), there can also be advantages to such non-belonging. In particular, the mobility and circulation partly made possible by flexible labour markets in the global city have both exploitative *and emancipatory* potential.

The mobility and invisibility offered by the urban and its 'low-skilled' industries to which (g)hosts are structurally constrained allows them to jump between employers and industries. Migrants are often ready and able to move for small improvements in conditions and pay (Forde and MacKenzie, 2009); of the (g)hosts surveyed by Wills and her colleagues, almost half had been with their current employer for less than a year (2010: 127). In her participant observation of migrant workers in London's hospitality sector, Gabriella Alberti found that taking temporary exploitative work could be 'tactical' or 'strategic', learning skills to 're-invest' in better jobs later (Alberti, 2014: 871–872; Cook et al., 2011). While assuredly vulnerable, (g)hosts need not always be docile and passive subjects of their host-fashioned precarity. Mobility between jobs and sectors thus becomes a possible 'terrain of resistance for low-income migrants' (Alberti, 2014: 866).

More than this, however, as global cities are the space of mobile transnational flows, cross-cutting and differential migrant trajectories, such flows also enable escape. (G)hosts need not be static, but are often following their own specific 'migration trajectories' (Alberti, 2014: 869) that do not have to end in a particular global city. The exploitative invisibility and non-belonging that constitutes their precarity and (g)host-ness can be embraced as one step on a longer migration facilitated by the open indifference of the city. Alberti found that some, such as Cynthia, an Eritrean with an Italian passport and Jamaican husband, made use of the flexibility of hospitality work to service family commitments while gaining skills and money which she would invest in a business in Jamaica. Felix, a Brazilian waiter, accepted the abuse he received for not taking his job seriously as it facilitated his journey to a better job and life in Paris (Alberti, 2014: 872–876). Others were in London just to 'explore', experience the world, and temporary work facilitated their travel plans.

Either way, (g)hosts may have no desire for the amnesties and regularisation of host-ness. They often have no wish to make London their 'home', to move from (g)hosts to hosts and therefore make little effort to 'integrate' with local communities (Wills et al., 2010; Batnitzky and McDowell, 2013). The abusive welcome London offers can sometimes be sufficient as a lay-over in their transnational lives; they make use of it, exploiting its openness and hostipitality just as it exploits them. For this reason, Mireille Rosello suggests that such subjects are not migrants, but 'transients'. Each has their own story, their own path and may require no 'welcome', just the opportunity to 'disappear' (2009: 24). Global cities, with their anonymity and indifference enable this. The global city is a home of mobility rather than stasis, its freeing ethos is one of '[d]welling as flowing'; transient tactics are oriented towards survival and removing the blockages on that flow (ibid.: 21). That trajectory or flow may be 'onward' to other locations (Alberti, 2014) or 'back' home (Wills et al., 2010: 136), but it does not seek regularisation and residency.

While these tactics of transience and mobility illustrate possibilities of (g)hosts' hospitable counter-conducts, they are severely limited. On the one hand, the freedom to move is restricted by a variety of factors, from family and class background, educational achievement, community support and transnational friendship networks, to race, class, gender and immigration status (Wills et al., 2010; Alberti, 2014: 873–5; Lewis et al., 2014). On the other hand, while such counter-conducts are a form of behaving other than the host perhaps intends, they neither severely hamper nor challenge the welcome provided by the global city. Indeed, by making use of this flow, they feed into and support the type of flourishing, efficient hospitality London offers. Without transiency, London could not welcome those guests it targets: the globally talented and wealthy. It suits global cities to enable (g)hosts to circulate, while placing sufficient blockages such that some will always get 'stuck' (Alberti, 2014: 873–874) and become docile precarious workers (Anderson, 2010). The transiency and mobility enabled by global cities illustrate how the behaviour of (g)hosts is both conducted and counter-conducted, even if often towards similar ends.

## HOSTING COUNTER-CONDUCTS: CITIES OF SANCTUARY

Nominal hosts as well as (g)hosts seek to redirect the global city's hospitality. One particularly influential practice of counter-hospitality can be found in the Sanctuary City (in the US), or City of Sanctuary (in the UK), movement. The tradition in Judaism of cities as spaces of sanctuary was central to Derrida's original interest in ethics as hospitality, as he argued that cities could act as an open refuge in a way nation-states cannot (Derrida, 2001). The idea's religious origins remain important in its modern incarnation: Christian denominations came together to declare the City of Berkeley, California a city of refuge for soldiers conscientiously objecting to their deployment to Vietnam in 1971 (Ridgley, 2013: 221–223). Likewise, religious groups were at the forefront of the sanctuary movement in the US in the 1980s that led to 20 cities and two states declaring themselves spaces of sanctuary for Central American refugees (Ridgley, 2008; Mancina, 2013). Amongst these explicitly or implicitly declared sanctuary cities were those marked by openness and tolerance, which Richard Florida (2005) argues are favoured by the creative classes: Portland, Cambridge, Madison, San Francisco and New York.

In the US, the Sanctuary City movement has explicitly challenged and countered the hostility towards irregular migrants offered by the Federal Immigration and Naturalization Service (INS) (Ridgley, 2008: 70). Many cities' sanctuary policies remained largely declarative, exhorting police, welfare and service providers to avoid unnecessary cooperation with the INS, fulfilling their roles without inquiring as to a subject's immigration status. San Francisco went further, embedding principles of hospitality, INS non-cooperation and status non-inquiry within the City's Administrative Code in the 1980s. This required city services to be provided 'to all city residents including undocumented refugees in an "immigration status-blind" manner' (Mancina, 2013: 206). San Francisco sought to officially governmentalise its hospitality as a counter to state-based hostility and deportation of the undocumented. The bureaucratic nature of this counter-conduct is highlighted by the revision of all forms used by the city to ensure that questions about immigration status were removed (Ridgley, 2013: 225). This coordination ensured that all authorities within the city would turn the same 'blind eye', allowing the space for charities and churches to shelter, feed and clothe refugees. The network of sanctuary cities has come under increasing pressure from Federal law-makers in the US, especially in the 1990s and post-9/11, forcing changes in San Francisco's policies. Most recently, Juan Francisco Lopez, an illegal Mexican immigrant with multiple felony convictions who had been deported five times before fatally shooting a woman on a pier in the city in July 2015, brought renewed national criticism (Philip, 2015). Nonetheless, the city's 1989 City of Refuge Ordinance remains in effect.

The US movement is partly enabled by the interaction between federal and municipal government, where freedom for local officials dwarfs that of more unitary states such as the UK. While Sheffield declared itself the first UK 'City of Sanctuary' in 2007, offering 'welcome and hospitality towards asylum seekers and refugees', this worked through local forms of connection and rebranding to create an 'alternative discourse' of welcome (Darling, 2010: 129–131). The aim was a more everyday transformation, building a 'culture of hospitality' from the ground up (Squire and Darling, 2013: 61). Far from the policed resistance of San Francisco, the subsequent movement 'operates as a fluid network of practices aimed at shifting hostile attitudes toward refugees and asylum seekers' (Bagelman, 2013: 50). There are still only seven recognised Cities of Sanctuary (Bradford, Bristol, Cardiff, Coventry, Newcastle, Swansea and Sheffield) in the UK. Despite a public meeting to gauge support in 2008, the formation of a draft constitution and the election of officers and a coordinating committee in 2009,[5] attempts to form a group to make London a City of Sanctuary flopped.[6] This culture of counter-hospitality has thus far failed to penetrate the UK's one global city.

The rigour of this resistance is also open to question. By accepting the difference between citizens and asylum seekers/refugees, the City of Sanctuary movement fails to challenge the firm statist distinction between belonging and non-belonging, guests and hosts (Squire and Darling, 2013). This is evident in its aim of developing 'a culture of welcome' so that 'people seeking sanctuary from violence and persecution in their own countries' can be given support and 'have their contribution to the community recognised and celebrated' (City of Sanctuary, undated). Reflecting statist understandings of 'deserving' or 'genuine' asylum seekers, sanctuary-seekers must have experienced persecution and make a contribution. Even San Francisco's sanctuary is a resistance to statist, sovereign power; it fails to challenge the flourishing hospitality already offered by the global city. In fact, sanctuary appears to work *with* the urban ethos, rather than producing an alternative, resistive form. After all, the global city's narrative of cosmopolitan diversity and liberating hospitality already counters the migrant-restricting narrative of the states they are embedded within but also exceed (see Massey, 2007). The city, unlike the INS in America or the Home Office in the UK, does not deport (g)hosts. So in what sense is urban sanctuary a counter-conducting hospitality?

Nicholas De Genova (2002: 438) argues that it is not actually deportation that renders (g)hosts a 'distinctly disposable commodity'; rather, it is their 'deportability', produced through the state's exercise of sovereign power. 'Deportability' encompasses asylum seekers, irregular migrants and those who have 'overstayed' or exceeded their visa.

> It is *deportability*, then, or the protracted possibility of being deported – along with the multiple vulnerabilities that this susceptibility for deportation engenders – that is the real effect of these policies and practices. *Deportation regimes are profoundly effective,* and quite efficiently so, exactly insofar as the grim spectacle of deportation of even just a few, coupled with the enduring everyday deportability of countless others (millions, in the case of the United States), that produces and maintains migrant 'illegality' as not merely an anomalous juridical status but also a practical, materially consequential, and deeply interiorized mode of being – and of being put in place. (Peutz and De Genova, 2010: 14 – emphasis in the original)

The global city does not generate the deportability of the (g)host through its hospitality, but it both *uses* it and *entrenches* it. It does so by depending upon the deportable subject, exploiting its labour and inability to claim rights to produce a cheap, efficient welcome. It ignores the deportable (g)host's lack of documentation (most of the time), but does not promise to always do so. It offers unsurveilled, cramped and squalid living-spaces to (g)hosts, so long as they turn up on time to clean, cook and secure the spaces of the desired guests. Crucially therefore, the (g)host is deportable; this is what prevents her 'belonging' in the city and keeps her as a (g)host: invisible, silent, hard-working and *easily governable*.

The global city's hospitality does not conduct the conduct of (g)hosts through deportation, then, but rather through galvanising their deportability. Sanctuary movements constitute a counter-discourse of hospitality by both challenging *and* reinforcing these tactics of government. Formalised and bureaucratised sanctuary practices, such as those in San Francisco, firm up the 'immigration blindness' of city service providers. In this sense, they mitigate deportability, ameliorating its worst effects by allowing (g)hosts to access social services, healthcare, public schools, police and fire service without fear of status-enquiries and deportation (Ridgley, 2008: 71; Mancina, 2013: 210). This reduces the vulnerability of the (g)host, potentially softening their exploitation and the worst excesses of violence experienced at the hands of employers and the INS. But sanctuary city hospitality does not challenge the deportation regime or the sovereign power of the INS. Rather, it simply withdraws the city's cooperation. No resolution is offered; just a more predictable and less vulnerable irregularity. Likewise, the educational endeavours of UK sanctuary cities build coalitions and aim to form connections between citizens and asylum seekers in schools, universities and museums. Their transformation of urban culture aims at *reducing* the non-belonging of (g)hosts, not exploding the distinction between guest and host.[7] City of Sanctuary offers some visibility to the (g)host, a space in which to articulate claims and rights with less fear. But this remains based on the host's understanding of being a worthy guest – fleeing violence/persecution and contributing to the community. In this sense, they offer (g)hosts a stage on which to *perform* and *prove* their worthiness. At best, such a performance will render them *less* deportable rather than non-deportable, easing their non-belonging.

As everyday changes to the ethos of a city, the sanctuary movement reduces the unevenness of urban hospitality and its mundane practices of inclusion and exclusion. It thereby makes the city a more welcoming, hospitable space. But the counter-conducting hospitality of sanctuary practices are easy to overstate. Jennifer Bagelman argues that its logic of amelioration also 'regularizes and depoliticizes' (Bagelman, 2013: 50). By limiting the worst excesses of the city's hospitality, sanctuary practices 'ease the seriousness of the problem' and 'incite a commitment to the rules of the game' (ibid.: 56). In my context, the rules of the game include commitment to the open and welcoming but exploitative and violent hospitality of the global city. The hosting counter-conduct of sanctuary moderates but also secures the global city's liberating hospitality, marginally empowering the (g)host rather than allowing her to flourish.

## CONCLUSION

When compared to the spaces examined in previous chapters, the global city's ethos of freedom and flourishing produces a markedly open home. Compared to the sovereign gaze of the hotel, the patriarchal pastoralism of the feminised family home, the invasive conditionality of the homeland (Chapter 1) and the caring restrictions of the refugee camp (Chapter 2), the global city's spectacularly indifferent cosmopolitanism appears attractive. Its many blind spots and non-totalisation also allows room for alternative ways to welcome and be received. Yet the genius of this flourishing hospitality lies in the way it generates the urban space and secures it through tactics of self-government which quietly nullify the threat posed by such resistance. Alternative ways of being a host or (g)host offer only minor empowerment and are readily taken up and subsumed by the city's management of freedom and constraint, flourishing and exploitation. Meanwhile, urban space is organised around catering to the desires of those guests it targets and neglects any form of responsibility for the guest-hosts, or (g)hosts, who do much of this 'catering' work. Thus, while the camp could afford to learn from the freedom of urban hospitality, so the city could attend to the more caring aspects of refugee camps' humanitarian welcome.

How can the city be 'bettered', then, as a space of hospitality? To approach this question, in necessarily tentative terms, we need to think both of the internal and external aspects of the post-sovereign ethical relational space. As Massey (2004: 6) notes, a great deal of work has focused on how to negotiate relations to the stranger *within* a space (e.g. Amin, 2012). In this vein, we can consider how the global city could place limits on the way it exploits (g)hosts – through regulation of the labour market, housing and social welfare provision – whilst avoiding such measures blocking the transient's ability to 'flow' through the city and onwards. Such policies would take much political and ethical imagination, but can only begin with a recognition of the responsibility owed to those who help constitute the global city and the welcome it affords. These policies could be oriented toward making the global city's flourishing hospitality more equally available. At best, however, such measures will mitigate the worst effects of the advanced liberalism which has become central to the urban ethos of flourishing. Perhaps we therefore need to rethink the ethical defensibility of the ethos itself.

In this concentration on the domestic aspect of the global urban home, however, we must not neglect the wider repercussions of the city's hospitality. As Massey notes, we must not stop at such 'an internal, indeed internalised, view of the city':

> It is about hospitality, about those who come to 'us', about the strangers within the gate … And this is excellent. However, the geographies of places aren't only about what lies within them. A richer geography of place acknowledges also the connections that run out from 'here': the trade-routes, investments, political and cultural influences; power relations of all sorts run out of here around the globe and link the fate of other places to what is done in London. This is the other geography – the 'external geography' of a place. (2006: 64)

It is these wider geographies of responsibility we must also take into account when considering the hospitality which constitutes the global city and has been ignored in successive London Plans. This would consider the effects on communities in Africa, Asia and South America (as well as other parts of the UK), whose doctors, nurses and other graduates become the guests of cities such as London, often becoming over-qualified (g)host's in their hospitality industries. London's hospitality thus generates responsibilities which could be accepted in various forms, negotiating with the communities around the globe harmed by its welcome, offering compensation and reparation (rather than aid) (Massey, 2006: 70). More radically, perhaps a responsible hospitality would require a targeted *refusal* to welcome the most talented and qualified from certain places, based on that particular community's needs. This is something EUrope (Chapter 5), attempts to deal with through 'ethical recruitment' outside certain sectors, whilst targeting 'brain circulation' rather than 'brain drain' (Council of the European Union, 2009: Article 5(3); Frattini, 2007a).

What is crucial to both the internal and external bettering of a global city's hospitality is that its free, flourishing ethos be tempered with greater care and responsibility. In the next chapter I turn to one of the spaces in relation to which London's status, accumulated resources and place in the global hierarchy has been built and secured: the postcolonial state. In particular, I offer a genealogy of the production of (Trans-)Jordan, a state which was both metaphorically and materially constructed through an almost unconditional hospitality towards its guests. By welcoming in outsiders, an apparently indistinct territory east of the Jordan river has gradually become a post-sovereign homeland defined by its hospitality to refugees and tourists.

## NOTES

1 For early discussions of the culture and ethos of the city, see Mumford (1938) and Wirth (1938); see Pile (1999) for a simple introduction and Amin and Thrift (2002) for a more complex discussion.
2 Illustrating the assemblage's temporary nature, this website has since been taken down. It was replaced by the much more complex 'map' (London Communications Agency, 2011), found initially at www.lse.ac.uk/geographyAndEnvironment/research/london/pdf/WhoRunsLondon_2010.pdf (accessed 25 February 2014). This has also subsequently been removed, but is archived by the LSE.
3 Available online at www.newlondonarchitecture.org/dls/whorunslondon.pdf (accessed 25 August 2014).
4 The other centres are 'Major centres', defined as retail spaces with a borough-wide catchment, and 'District centres' which provide 'convenience goods and services for more local communities' (Mayor of London, 2014a: 317).
5 Information was formerly available online (www.cityofsanctuary.org/content/about-london-city-sanctuary – accessed 17 October 2014). However, the link has subsequently been removed from the City of Sanctuary website.
6 Emailed requests for information sent to the address included on the City of Sanctuary website went unanswered.
7 Though it can offer the space for 'minor' acts which do precisely this (see Squire and Darling, 2013: 69–70).

4

# Unconditional Hospitality: (Trans-)Jordan as Postcolonial State

The space of contemporary Jordan is one marked over and over again by a history of hospitality. This has been exhibited in recent years through the welcome offered to Syrian refugees. Speaking to the European Parliament in March 2015, reigning monarch, King Abdullah II, underlined the morality and extent of their generosity:

> Jordan also takes seriously our moral obligations to others. Despite scarce resources, the people of Jordan have opened their arms to refugees fleeing regional violence. Jordan has taken in thousands of Iraqi Christians over the past year. This is in addition to giving shelter to 1.4 million Syrian refugees, which is 20 per cent of the population, over the past few years. This is more than the equivalent of France hosting the entire population of Belgium. My small country is now the world's third-largest refugee host and I thank all of you who are helping us to uphold this global responsibility. (King Abdullah II, 2015)

These figures do not match those of the UNHCR (2015: 3, 12), which places Jordan sixth in global refugee hosting, yet they demonstrate the scale of Jordan's efforts and their couching in terms of morality. This hospitality has been singled out for praise by the EU (Avramopoulos, 2014b) and humanitarian NGOs such as CARE International (Mitscherlich, 2013) and MercyCorps (Mason, 2013). It has been described by the Brookings Institution as 'phenomenal' (Bradley, 2013). The International Peace Institute's *Global Observatory* places Jordan's hospitality within the common narrative of welcoming over two million Palestinian refugees and many thousands of Iraqi refugees during the two Gulf Wars, such that now 'Jordan is a nation of refugees' (Christopherson, 2015). Palestinian refugees in particular now constitute over half its population. Some even suggest that Jordan approaches Derrida's limit-testing conception of 'unconditional hospitality' (Bell, 2014).

This is to overstate the case. Human Rights Watch (2014: 564, 2015: 322) draws attention to the various restrictions Jordan has placed on the entry of Syrians, Palestinians and Iraqis, particularly unmarried men of fighting age and those without documents. It is also to misunderstand the possibility and desirability of such an unconditional hospitality, as outlined in the Introduction to this book. Yet there is something significant in the range of praise Jordan receives for its outstanding hospitality. What it often enshrines is an 'ethnographic habit' of understanding Mediterranean and Middle Eastern societies as 'distinctive in their commitment to hospitality' (Shryock, 2012: S21). Ethnographers have done a great deal to disentangle the ways in which hospitality – or *karam* (hospitality, generosity, nobility) – are woven throughout the changing tribal social and political structure of Jordan (see Layne, 1994; Shryock and Howell, 2001; Shryock, 2004, 2009, 2012). Yet none has revealed the way in which hospitality and the power relations between hosts and migrant guests actually produced the very space and ethos of the postcolonial state of 'Jordan'. Emerging from Ottoman suzerainty, the *almost* unconditional welcoming of two foreign 'guests' in 1920 – the British, as the state responsible for the post World War I Mandates over Palestine and Iraq, and the Hashemites of the Hijaz – resulted in a power struggle that generated the boundaries and culture of this postcolonial state.

This chapter offers a brief and disjointed genealogy of hospitality's role in the formation of (Trans-)Jordan as a way of revealing the contingency of its current international identity as a state which welcomes refugees and tourists. Despite the various 'styles' of genealogical inquiry, all are united in 'the task of denaturalizing objects and subjects, identities and practices that otherwise appear given to us, lessening the stranglehold they exert on our political imagination' (Walters, 2012: 118).[1] As a genealogy, the chapter is not concerned to capture what is 'pure' or 'essential' (Foucault, 1991b: 78) – neither the essential nature of 'Jordan' nor its hospitality. Rather, it explores how the accidental and non-determinative openness of (Trans-)Jordan generated a 'play of dominations' or 'struggle of forces' (ibid.: 83–85) between the different ethical projects of 'good hosts' and 'bad guests'. It was this struggle that brought the space into being as contested, insecure and always in the process of becoming something else. The chapter then explores how this clash galvanised the borders of (Trans-)Jordan, firming it up as a *particular* space – a postcolonial or quasi-state – through a set of adaptive disciplinary power relations that pacified the desert and the irruptive ethos of the Bedouin. The British and Hashemite guests destroyed and then reconstructed a Bedouin ethos that could safely bound a sense of national identity – 'Jordanianness' – which would see the British finally ejected from its 'house politics'.

This 'struggle of forces' is traced through the memoirs and autobiographies of Jordan's guests, in particular those of the Hashemite Kings – King Abdullah I[2] and King Hussein[3] – and those of the British – the prolific John Glubb (Commander of the Desert Patrol and Arab Legion)[4] and Herbert Samuel (British High Commissioner of Palestine).[5] The final section shifts to more recent events, examining the way 'traditions' of hospitality have been commodified with the aid of new guests (United States Agency of International Development – USAID) in order to sell Jordan to the international tourist market. These externally driven adaptations and 'improvements' of Bedouin *karam* now help define Jordan's way of being as an international actor. In so doing, they have generated resistances and reveal fault lines that USAID can do little to discipline or efface. First, however, I want to situate Jordan within IR debates as a 'postcolonial state', defined by openness and powerlessness – more 'pre-sovereign' than 'post-sovereign'.

## (TRANS-)JORDAN AS A POSTCOLONIAL STATE?

(Trans-)Jordan[6] is unlike the spaces previously examined in this book, where power is initially exercised by a host as (often an assembled) sovereign master. Whether that power operates through the decision to welcome the guest and on what terms, the prefiguration of the guest as a population, or targeting specific guests, the encounter is unequally balanced in favour of the host. Once the threshold of the home is crossed, various power relations overturn and blur these distinctions, but in the early encounter, it appears to be the host exerting most power. Yet with (Trans-)Jordan, the opposite seems to be the case. Here it is the guests – the British and Hashemites – who are predominant; the host appears disparate and relatively powerless. The host and its home are caught between colonisation and postcolonisation: the Ottoman Empire and the British Mandate; the British Mandate and Hashemite Arab Nationalism; independence and sovereignty.

Following the successful conclusion of the 1916 Arab Revolt against Ottoman rule declared by Abdullah I's father, Sharif Husayn bin Talal of the Hijaz – Jordan's 'founding myth' (Alon, 2009: 2) – the Hashemites fought alongside their British allies against the Ottomans during World War I. Husayn then pronounced himself King of the Arabs, aiming to unite what the Hashemites understood as 'natural Syria' (King Abdallah, 1978: 27; King Hussein, 1978: vii–viii), the Arab homeland,[7] installing his sons as leaders in Syria (Faisal) and Iraq (Abdullah). This plan was hampered by the victorious European powers, Britain and France, and their strategic interests in the region. Their secret negotiation of the Sykes–Picot agreement in 1916, contravening assurances Britain had given to Husayn, divided this 'natural' Arab homeland into areas of French (Lebanon and Syria) and British (Iraq and Palestine) control. To this end, at the Paris Peace Conference in 1919 France and Britain agreed to make cynical use of the League of Nations' mandate provision, established in Article 22 of the League's Covenant.

Syria, Palestine and Iraq were thus placed under the 'tutelage' of Britain and France until they were ready to 'stand by themselves' as modern, sovereign states. While Faisal continued to administer the British-controlled area of Syria, the British withdrawal of its forces and French military takeover a few months later cast him out (Wilson, 1987: 39–43). Faisal had made little attempt to govern the space which became Transjordan, but with Faisal gone it 'became in essence a no man's land whose 300,000 peasants and semi-nomads were ruled by local sheikhs who did as they pleased' (Sicker, 2001: 61). Thus, what Joseph Massad (2001: 11) calls Transjordan's 'colonial moment, its very inaugural moment', came in 1921 with its establishment as an Emirate under the British Mandate for Palestine, with its own 'autonomous' rule by Hashemite Amir Abdullah (and a British Representative). The British and Abdullah would struggle to establish effective control, over each other and over the territory and population of Transjordan for the next 26 years, before its formal 'independence' and 'sovereignty' as a kingdom were granted by Britain in 1946, becoming the Hashemite Kingdom of Jordan in 1949 (see Wilson, 1987; Massad, 2001; Alon, 2009). Even after it formally achieved 'sovereignty', the British subsidised Jordan through an annual grant and controlled large parts of its armed forces until Glubb was dismissed as Commander of the Arab Legion in 1956. Even now Jordan relies upon the US to underwrite its security (Sharp, 2010) and depends upon substantial flows of international aid, over $1.4 billion in 2013 alone (World Bank, 2015), to keep its economy afloat.

It is therefore difficult to conceive of (Trans-)Jordan throughout its 94-year history as a sovereign state or space in traditional IR terms. Conventionally, IR has treated states as 'like units', sovereign actors that differ only in their power capabilities (Waltz, 1979: 91–99). However, more liberal IR theorists came to see such a vision as too reductive and incapable of accounting for the different functional capacities of 'sovereign' states in the global north and south (see Bull and Watson, 1984; Jackson, 1987, 1990; Sørensen, 1997). Robert Jackson therefore differentiated between a 'real state' and a 'quasi-state' (Jackson, 1987: 542): while real states existed in the developed West and Soviet Bloc, he identified quasi-states with the 'third world'. Jackson does not mention Jordan in this regard, or any Middle Eastern state, locating quasi-states almost exclusively in postcolonial Africa where national boundaries are artificial, central state authority is absent, corruption is endemic and public realms are 'fairly loose patchworks of plural allegiances and identities somewhat reminiscent of mediaeval Europe' (ibid.: 526–528). Quasi-states are granted juridical sovereignty only by the 'courtesy' of international society, which enshrines their right to self-determination. Quasi-states possess only 'negative sovereignty', derived from this right and courtesy, 'without yet possessing much in the way of empirical statehood, disclosed by a capacity for effective and civil government – positive sovereignty' (ibid.: 529).

Georg Sørensen (1997) developed this analysis, using Weberian ideal-type methods to argue that there are now three varieties of states cohabiting in the international system: 'Westphalian', postcolonial and postmodern.[8] The latter only really applies to Europe, where states have pooled sovereignty in the EU as a supranational institution (Sørensen, 1997: 261–264). Westphalian states (Jackson's 'real states') possess a 'substantial' form of sovereignty by having the 'capacity for self-government', a self-sustainable economic resource base and the ability to both maintain domestic order and military defence. They are bound together by a 'we-ness' which may be based in common cultural or ethnic roots (ibid.: 258–259). In contrast, postcolonial states, echoing Jackson, have only negative sovereignty due to their incapacity for self-government, 'weak and underdeveloped' state institutions, systems of personal rule, aid-dependent economies and lack of binding 'we-ness' (ibid.: 260–261). Postcolonial states suffer from internal rather than external threats to security – they 'are not internally pacified' (ibid.: 264). While in the past 'such weak states would have gone under', they are now 'forced to take what they can get from richer and stronger countries in the North', with their policies determined by the conditions placed on aid by donors (ibid.: 264–265). Again, Middle Eastern states remain unmentioned in their fit with Sørensen's ideal types; Sub-Saharan Africa most closely fits the postcolonial ideal type.

There are major problems and dangers associated with these analyses. At best, they support the colonial logic of dominance through their implication that Westphalian and postmodern states 'tutor' postcolonial states until they are capable of a more 'substantial' form of sovereignty (Doty, 1996: 147–156). As we shall see, this is precisely what Britain claimed to be doing in their administration of Transjordan. At worst, they justify forms of violent intervention and state-building; their reasoning is very closely reproduced in foreign policy discourses that advocate military interventions in 'failing states' (Bulley, 2008, 2009).[9] Theoretically, both models of differentiated sovereignty operate by making the 'Westphalian' or 'real' state appear natural – other types are derivative, defined only in relation to how they differ from the 'real'. As such, the differences between systems of government and rule in postcolonial states become irrelevant; what matters is that they do not live up to the requirements of 'substantial' sovereignty. Likewise, potential similarities between the postcolonial and postmodern are irrelevant to that which stands between them – the Westphalian state.

The absence of (Trans-)Jordan – or any Middle Eastern state – in these analyses also reveals the difficulty of placing its slipperiness, its perpetually 'Trans' identity, in the sense that it is 'crossing' or 'becoming' other. A mere Mandate Emirate in 1921, the state-building project of the British and Hashemite guests have since constructed a surprisingly stable but complex 'hybrid' state (Alon, 2009: 151) where power, sovereignty and identity are caught in a perpetual negotiation between ideas of family, tribe and state (Layne, 1994; Alon, 2009). Such negotiations take place, to a large extent, through traditions, practices and memories of hospitality, as we shall see below. While ideal-type methodology is necessarily reductive, analysing forms of power on the basis of a Eurocentric ideal of the state inevitably generates problems accounting for a space such as (Trans-)Jordan across time. Despite its current stability, central authority and domestic order, it remains for many an entirely artificial construct, produced by migrants:

> 'Outsiders' conceived of its borders and identity; they led its national army well after independence; people whose roots within existing memory lie outside the new borders of the country, ruled and continue to rule it; its population consists in its majority of people whose geographic 'origins' within living memory are located outside the borders of the nation-state ...; the country has a large dependence on foreign money to support its resource-poor economy; and claims are put forth by neighboring powerful states on its very identity (Israel, Saudi Arabia, and Nasirist Egypt, to list the more prominent ones historically), or on parts of it (the West Bank and Palestinian Jordanians) by a strong nationalist movement (namely, the PLO). It is in the context of this wide array of factors that Jordanian nationalist discourse has a more difficult time stabilizing the terms and essences it posits than the nationalist discourses of other postcolonial nation-states. (Massad, 2001: 15)

However, it is precisely this openness, ontological instability and unfinishedness that is encapsulated by its continued 'Trans-' identity, making it so hospitable and interesting to an ethics of post-sovereignty. The next section investigates the role that hospitality played in the encounter between the ethical cultures of hosts and migrant guests which began the process of producing a particular space out of a seeming 'no man's land'.

## THE HOSPITALITY OF (TRANS-)JORDAN: GOOD HOSTS AND BAD GUESTS

The characterisation of (Trans-)Jordan as a 'no man's land' in 1920 is deeply misleading – it was occupied by tribes, both sedentary and nomadic. The Hashemites had their own interpretation of the space; Hussein would later call it the 'remnants of the Arab homeland' which Abdullah was able to 'salvage' (King Hussein, 1978: vii). It was only conceivable as 'no man's' from a (colonial) perspective which defined territory according to its government – a view replicated in the classification of sovereign states seen above. The British saw it as 'the wild and unwanted territory east of the Jordan' (Glubb, 1948: 58), a mere rump remainder of the mandate carve-up, with Syria to the north (French), Palestine to the west (British), Iraq (British) to the east, the independent Kingdom of the Hijaz (Hashemite) to the south and the Nejd deserts (partially controlled by Ibn Saud) to the south-east. The space's absence of recognised authority had been its identifying feature since Ottoman rule; once Faisal's 'precarious control' was gone, the tribes of this area 'commenced with considerable zest the process of doing exactly as they felt inclined' (Glubb, 1948: 55–57). As High Commissioner for Palestine, this constituted a 'problem' for Herbert Samuel, a 'serious one': it was included within his Commission and the French dispossession of Faisal meant that the space was 'left in the air' (Samuel, 1945: 159).

Britain's interpretation of the space as 'wild', 'ungoverned' and 'in the air' was highly productive. It provided a post-hoc justification for intervention, for which there was initially no appetite in Whitehall (Wilson, 1987: 44–48), constituting the area as entirely open and welcoming to the entry of the British. After all, it was empty, the complete blank slate onto which Britain could write its state-building project. Whilst Samuel clearly relished his task of 'lay[ing] the foundations of a new State' in Palestine, the past of which was a 'panorama of civilisations' (1945: 156), he had little interest in its wild hinterland. It was only when Samuel believed the British to be *invited* into the 'country east of the Jordan' that he envisaged active incursion. As he described to King George V,

> The most interesting development during the last two months has been the visit of a number of sheikhs of the country to the east of the Jordan, to ask for their separation from the Government of Damascus, now that it is under French control, and their definite inclusion within the British sphere. I must have received, at various times, fully one hundred of these sheikhs, all of whom came with the same request. As a consequence, at the end of August, I rode over to Salt … and held there a meeting attended by over 600 representatives of the tribes and of the town population. *The feeling was unanimously in favour of British assistance in the administration of the country … The people are continually asking for the presence of British troops in the area,* and it is a somewhat striking contrast between the conditions north of the Sykes–Picot line, where the French are experiencing bitter resistance to their demands to enforce authority with large forces, and the conditions south of the line, where the whole administration is controlled by the people themselves, guided by six British officers, and *where continual demand is expressed for the presence of British troops,* which it is, however, not in accordance with our present policy to send. (Samuel, 1945: 159–160, emphasis added)

For Samuel, Britain's entry into this space was not at its own instigation. It was considered a hospitality of *invitation* rather than visitation (Derrida, 2003: 128–9). This separates it from the violence of conventional colonialism, where the guest appears without invitation and violently seizes the home of the host (Baker, 2011), an abuse of hospitality railed against by Kant when determining the rights of guests (Kant, 1991; Brown, 2010). British officers were certainly 'guests', as they did not *belong* there in any sense – it was only 'through events strange and unforeseen' that a 'people from a far-away island in the North Sea' had taken on this 'task' (Samuel, 1945: 146). Yet, accepting an invitation means also accepting its terms, which include the sovereign mastery of the host who invites (Derrida, 2003: 128). The British would prove to be bad guests, uninterested in the terms of their invitation.

Samuel's claim that invited British officers were to be mere 'guides' to an administration 'controlled by the people themselves' was belied by the 'fresh complication' that arose a few months later: the arrival of Abdullah in Ma'an, on his way through Transjordan to attack the French in Damascus (Samuel, 1945: 160). There was, however, a 'simple solution': 'Let Abdulla, then, give up his campaign against the French and settle down in Transjordan; let him be recognized as ruler there and be given the help of a few British advisers, and the moderate subsidy essential for a proper government' (ibid.: 160). The solution was for the British, themselves guests, to give away a space in which they did not belong to *another* guest.

A further complication, unimagined by Samuel, was that he may have misinterpreted his 'invitation'. The tribal leaders who he met with at Salt were not necessarily the most significant or influential (for instance, no representatives of the Bani Sakhr were there (Alon, 2009: 22). More importantly, while Samuel perceived the friendliness of tribal leaders as offering unanimous support for a substantial British presence, this was a misunderstanding based on the 'huge cultural gap' between the host and guest: 'the Arab etiquette that demands perfect hospitality, respect for guests and extreme politeness, does not necessarily imply political agreement or support' (Alon, 2009: 22–23). The British misjudged their status as guests and ignored any terms of their 'invitation'. As the British and Hashemite guests converged on the hosts of this open territory, what occurred was a clash of cultures, a collision of *ethea*, ways of being and belonging within a space that appeared wild, ungoverned and welcoming. This was to be a conflict of good hosts and bad guests that would create the space as (Trans-)Jordan.

## GOOD HOSTS: ARAB AND BEDOUIN HOSPITALITY

Glubb follows a tradition of nineteenth-century travel writing in identifying the Arabs and Bedouin of Central Arabia with the ethical practice of hospitality (Shryock, 2008: 412–413). Using 'Arab', 'Bedouin' and 'Transjordanian' interchangeably, his memoirs overflow with stories and descriptions of how the 'Arab tent' acts as a 'sanctuary not only for himself [the host] but for all the world who may appeal to it' (Glubb, 1948: 136; see also Glubb, 1948: 43, 78–79, 119, 137–139, 155–157, 189 Glubb, 1957: 370–372; Glubb, 1971: 2–3; Glubb, 1983: 59, 66, 100). 'In one quality, the Arabs lead the world – it is the virtue of hospitality' (Glubb, 1957: 37). The fact that this description comes after a rendering of the essential characteristics of various peoples, from the 'lethargical' Egyptians to the 'virile race' in Central and Saudi Arabia (ibid.: 32–37), alerts us to the Orientalism of Glubb's observations.

For Andrew Shryock, an anthropologist who spent much of his career studying the politics of hospitality amongst the Balga Bedouin of Jordan, such Orientalism does not mean that these observations can be dismissed (Shryock, 2012: 21).[10] He argues that, for the Bedouin, hospitality is a culture, a way of being and dwelling in a space that is shot through with power relations regarding the reception of the guest:

> Hospitality, *karam* in local dialect, is not simply a matter of offering tea, cigarettes, and pleasant conversation to guests. It is also a test of sovereignty. The man who is *karim* (hospitable, generous, noble) is able to feed others, project an honourable and enviable reputation, and protect guests from harm. Hospitality, as Bedouin describe it, is a quality of persons and households, of tribal and ethnic groups, and even of nation-states. At any of these levels of significance, failure to provide *karam* suggests low character and weakness, qualities that attract moral criticism … This negative potential makes bad hosts and bad guests important. It gives moral focus to the central problem of hospitality: namely, how to enact autonomy and exchange, openness and closure, within the same social space. (Shryock, 2012: S20)

There are, as Shryock (2008, 2009) notes, strong similarities between the Bedouin ethos of hospitality and Derrida's (2003: 128) concept of 'pure and unconditional hospitality, hospitality *itself*'. This is the hospitality of the *visitation* rather than invitation, that which 'opens or is in advance open to someone who is neither expected nor invited, to whomever arrives as an absolute *visitor*, as a new *arrival*, nonidentifiable and unforeseeable, in short, wholly other' (Derrida, 2003: 128–129). Unconditional hospitality must occur without asking of the visitor 'either reciprocity (entering into a pact) or even their names' (Derrida, 2000: 25). Likewise, Bedouin Arabs make no pact, though reciprocity is assumed as part of a shared culture, and 'through the ages, have hosted a stranger for three days and a third before asking his name' (Fawzi al-Khatalin in Shryock, 2008: 407). After this period has passed, not only can the name be asked, the guest can also be cast out. Even in its idealised form, Bedouin hospitality is *not* unconditional, though the practice approaches it more than those examined in previous chapters.

Whilst Bedouin hospitality is 'fantastic, dramatic, romantic and unconnected with the practical needs of the situation' (Glubb, 1948: 155) such that it 'is often embarrassing' (Glubb, 1957: 370) and can push the host family towards starvation (Glubb, 1948: 43), it is also a common saying that the guest 'is the prisoner of the host' (Shryock, 2012: 23). Various restrictions are placed on guests. Visitors are not allowed free reign over the Bedouin home as would be required in an unconditional hosting (Derrida, 2000: 77) – they are limited to the space set aside for entertainment (Shryock, 2012: 24).

In the desert, this will often be the male half of the tent, separated by a hanging blanket (Layne, 1994). The gendering of Bedouin hospitality imposes a strict condition: it is something men offer to other men. Like the (g)hosts of the global city, Bedouin women may do the work of *producing* the food and drink and, as property of the host, can be offered as gifts to the guest (Shryock, 2008). Female guests are entertained separately in a less lavish manner (Shryock and Howell, 2001). As prisoners, guests cannot leave until they have eaten, using *only* the best of the utensils (Shryock, 2012: 24). Bedouin hospitality's very excess is itself a condition: its performance demonstrates the host's mastery of the space and the guest's non-belonging. Its disproportion is as much about control as more limited hospitalities. It is the 'proof and practice of sovereignty' (Shryock, 2004: 58).

Bedouin hospitality can therefore appear close to the Western 'tolerance' which Derrida sees as the limit of its conditional form. Tolerance welcomes on the basis of retaining sovereignty and keeping otherness in its (powerless) place (Derrida, 2003: 127–128). The difference is that Bedouin practices explicitly *retain* an element of danger which tolerance guards against. True *karam* 'creates, in guest as well as host, a tantalizing sense of *risk*' (Shryock, 2004: 40), as its excessiveness 'expose house and host to danger' (Shryock, 2008: 415). The Bedouin proverb continues, 'The guest is prisoner of the host', but 'The host must fear the guest. When he sits [and eats your food] he is company; when he stands [and leaves your house], he is a poet' (Shryock, 2008: 415, 2004: 36). Once the guest is released, he is free and can make or break the host's reputation for *karam*, telling stories which destroy his sovereignty, his reputation for a generous, noble, *hospitable* way of being. *Karam* then – encompassing sovereignty as true, noble mastery and hospitality – necessitates danger; both host and guest are exposed, negotiating their vulnerability and power in the practice and aftermath of welcome. This is not sovereignty as complete control of violence; rather it is sovereignty as unguardable exposure. 'Without this sense of risk, hospitality loses its moral power' (Shryock, 2008: 415).

Shryock (2012) shows that shifts in political power are narrated in Jordanian history through stories of bad hosts and bad guests – power is exercised through practices which *break* with Bedouin tradition, turning this vulnerable sovereignty against itself. Thus, perhaps the subjugated histories of the encounter and struggle between the Central Arabian tribes, British and Hashemites can be understood in this way – as a hospitable relation between *good* hosts and bad guests, with sovereignty as the elusive prize. To call these tribes 'good hosts' is not to romanticise their actions; it merely stresses that they welcomed in keeping with an ethos of vulnerable sovereign hosting. The hospitable ethos of the territory and people East of Jordan may have been misinterpreted by the British guests who outstayed their welcome and gave away their home. The Hashemites, the bad guests who received this gift and subsequently ousted the British, are still there today; they never left. Indeed, King Abdullah II, constitutional monarch since 1999, symbolises the confluence and hybridity of archetypal bad guests: the son of King Hussein and his British second wife, Princess Muna Al-Hussein (formerly Toni Gardiner).

## BAD GUESTS: BRITISH COLONIALISM AND HASHEMITE ARAB NATIONALISM

In contrast to the Bedouin ethos, the *ethea* of the British and Hashemites demonstrated a very different understanding of the relationship between host and guest. The British way of being and dwelling as a guest was, in their own terms, one of *service*. For Glubb the 'moral basis' of empire was 'founded not on domination of other peoples but on serving them' (1983: 213). Britain's role in its empire was simply 'to serve the other members'. Finding themselves at the borders of a no man's land, because of events 'strange and unforeseen', what could they do but be of service? For Samuel, the region contained a population whose 'great majority were eager to be freed from an alien [Ottoman] rule that had been marked by gross misgovernment' (Samuel, 1945: 139). With British help, 'a different future would then be opened for these provinces'. The genius of Transjordan, and Abdullah in particular, was in recognising Britain's 'disinterested and devoted service to our Eastern brothers' (Glubb, 1948: 221). After all, 'The Arabs could not progress without European help, yet they feared to accept that help lest it lead to European domination'. Those that refused this 'help' remained 'backward and chaotic. In Jordan alone, under the influence of King Abdulla, was the assistance whole-heartedly accepted, yet without submission ... always remaining an equal' (Glubb, 1957: 445).

This 'disinterested' ethos of service is belied by both Glubb and Samuel acknowledging that British interests where paramount in the help offered. Such 'strategic interests' included preventing Palestine, bordering the Suez Canal, falling 'under the control of any of the great Continental powers' (Samuel, 1945: 139–140). Glubb massages this cognitive dissonance by clarifying that 'Britain had no interest in the Arab countries themselves'; rather, Britain's interest was in these countries as a highway, a road to elsewhere (India, China, Malaya). It was this 'right of transit' that was their interest (Glubb, 1957: 375). Such a fine distinction bears little scrutiny. Nontheless, it is significant that Glubb approvingly cites Palmerston's claim that 'we do not want Egypt or wish it for ourselves' just as a 'rational man' with property in the North and South of England would not wish 'to possess the inns on the North Road'. Reflecting Britain's view of the region as one from which they required (and demanded) hospitality, all the rational man 'could want would have been that the inns should be well kept, always accessible, and furnishing him when he came, with mutton chops and post horses' (Palmerston in Glubb, 1957: 375). The ethos of the British guest was one of self-interested service to the host, securing continued hospitality and service in return. This was best provided by the hospitality of the Hashemites rather than a no man's land.

In contrast, the ethos of the Hashemites was that of a guest who explicitly saw themselves as sovereign hosts, though they also couched this in the language of 'service' and 'sacrifice'. Their wider cause was that of Arab nationalism (King Abdallah, 1978: 4). This Arab nation is 'a people formed in the mould of absolute freedom; one might call this the freedom of the Bedouin, which is basically the liberty to do as one likes without restriction', though Islam served to 'bound the Arabs with its traditions and way of life and prepared them to serve humanity' (King Abdallah, 1978: 5). However, with Faisal cast out of Syria in July 1920, the aim of uniting 'historic' and 'natural' Syria – 'a pillar of the Arab community' (King Abdullah, 1950: 191) – was in tatters. Thus, when Abdullah set out with an army from Mecca in the autumn of 1920, arriving at Ma'an in November, it was not to become Amir or King of Transjordan. As he proclaimed to Syrian Arabs from Ma'an, his aim was to 'drive the [French] aggressors from our shores', such that they 'realize that the Arabs are one body; if one member suffers the whole is affected' (ibid.: 191).

This was the 'problem' that the British were confronted by in the winter of 1920: allowing 'the British mandatory area to be used as a base for an attack on our French neighbours would never do ... [but] we did not want to be led into a clash with our Arab friends' (Samuel, 1945: 160). Abdullah indicates that local chiefs, like the British, did not welcome him. The Governor of Salt wrote telling him that they had heard of his 'intention to stay in Transjordan. If your visit is a private one, the country will welcome you', but if it is political they would 'do all to stop your coming' (Mazhar Bey Raslan in King Abdullah, 1950: 193). Abdullah's reply was not that of a liberator – 'I am visiting Transjordan to occupy it'. Britain, assuming sovereign mastery of what was merely a mandatory area, had a conflict of loyalties regarding its 'neighbours' (French Syria) and 'friends' (Hashemites), but ignored the Transjordanian hosts' desires. Its 'simple solution' has been described and the liberating-occupier Abdullah arrived in Amman to become Amir in March 1921. The Hashemites' non-belonging in (Trans-)Jordan is acknowledged by Abdullah, who proudly declares his 'own country' to be 'the most sacred Hejaz' (King Abdullah, 1950: 37). His service included sacrificing his home in the Hijaz (King Abdallah, 1978: 31–32), which was conquered by Ibn Saud, with King Husayn exiled to Cyprus after Britain prevented Abdullah offering him sanctuary in Transjordan (Wilson, 1987: 88).

What we see in the early 1920s is a complex hospitality where the welcome of the (Trans-)Jordanian space and its ethos of (almost) unconditional hospitality led to its subjugation by British and Hashemite guests. While both guests claimed an ethos of service and liberation from colonial rule (the Ottomans for the British; the French for the Hashemites), strategic interests were served in all three cases. The welcome of the hosts was a failed negotiation of a vulnerable sovereignty; the British sought to preserve the openness, pliability and hospitality of the space to service their own mobility; the Hashemites sought the unification and liberation of an Arab nation with themselves as unfettered sovereigns. Each ethos involved governing and resisting the others in this emergent space. The result was initially most favourable to British interests, but the problem remained as to what to make of the space and its people. After all, before 1921 'there was no territory, people, or nationalist movement that was designated, or that designated itself, as Transjordanian' (Massad, 2001: 11). The task of the guests was literally to *produce* a territory, people and national 'we-ness' from this space – and 'to create rather than to take over the Government' (Glubb, 1948: 58).

## THE GUESTS' PRODUCTION OF (TRANS-)JORDAN

According to Abdullah, the creation of the (Trans-)Jordanian state was miraculously completed through his presence. 'After my entry into Amman, the whole of the Transjordan was occupied and orders were sent out from Amman' (King Abdullah, 1950: 200). In reality, producing a state was not this simple: a territory comparable to that of England, five-sixths of which was desert, including the seemingly impenetrable Jabal Tubaiq mountains to the south and the volcanic Jebel Druze to the north (Glubb, 1948: 85, 108–112), could not be occupied and controlled by a handful of British officials and the small army brought from the Hijaz. Samuel appears to endorse Abdullah's line; more likely, he was relieved to relinquish the 'cares and the risks' of directly governing such a large territory with poor communications 'and backward conditions'. A further advantage was that 'the traditional Beduin [*sic*] raids into Palestine could be stopped at their source by an Arab prince well accustomed to handling the Bedu tribes' (Samuel, 1945: 161).

The practice of raiding appears here because it was central to British understandings of the territory as wild and ungoverned. The nomadic and semi-nomadic Bedouin tribes had no respect for orders sent out from Amman, nor the recently invented national boundaries produced by colonial rulers of whom they had never heard. Tribal practices of raiding, both within Transjordan and across its 'borders' with Palestine, Syria and the Nejd, killing their enemies and carrying away camels, sheep and other goods demonstrated precisely the characteristics of the 'postcolonial state' identified in IR – '[s]uch states are not internally pacified' (Sørensen, 1997: 264). Both the deserts and towns of Transjordan were constituted by 'fairly loose patchworks of plural allegiances and identities' (Jackson, 1987: 528) of Circassians, Kurdish, Armenians, Chechens, Hijazis, Syrians, Palestinians and the many 'native' tribes, with no 'developed nationhood' or sense of 'we-ness' (ibid.: 261).

Samuel's rosy view of Abdullah's capabilities as an 'Arab prince' was plainly ridiculous given his own outsider status and lack of authority over the deserts. The power struggles between the British and Abdullah over how Transjordan would be managed, with their very different conceptions of the space (as a 'buffer' for Palestine and highway to the East; as a stepping stone to Syria) through the 1920s are charted by Mary Wilson (1987: 60–102). However, as it became apparent that Transjordan could not be sustained as a mere rump and would have to become more state-like, the necessity of pacification and the generation of a national identity became paramount. This section examines how an imperial, patriarchal improvisation of disciplinary power was exercised by the British and Hashemites to firm up the home within which neither belonged. In pacifying the borders and building a 'native' military, Glubb played a central role from 1930 onwards. The Hashemites, feeding from Glubb's successes, concentrated on negotiating a workable sense of 'we-ness', building the affective belonging necessary to a home.

## PRODUCING/PACIFYING SPACE

Glubb, or 'Glubb Pasha' as he became known, arrived in 1930, having spent the previous decade in mandate Iraq, ending the cross-border raiding between Iraq and Nejd tribes, particularly the feared Saudi *Ikhwan*. Such pacification was deemed necessary due to British commerce and mobility in the region; British interests could no longer be left at the 'mercy of raiders and robbers' (Glubb, 1948: 72). As Mark Neocleous (2011) argues, capital accumulation is always central to pacification. Yet, with the official end of the mandate in 1928, Transjordan had 'taken no steps to enforce its authority in its deserts' (Glubb, 1983: 95). Hearing of his success establishing order in Iraq, the British Representative asked him to do the same in Transjordan (Glubb, 1983: 95). Britain remained officially responsible for the external defence of Transjordan and securing its borders, but had experienced little success in preventing cross-border raiding by Transjordanian Bedouin or Saudi *Ikhwan*.

Glubb therefore arrived to end the raiding that was destabilising the imaginary space of Transjordan – beginning the move from a 'postcolonial' to a 'Westphalian' state. His role was to 'introduce some order to this chaos' by 'tak[ing] over control of their desert' (Glubb, 1983: 99). The methods Glubb employed would not involve the *physical* destruction and reconstruction identified by Neocleous (2011) in colonial practices of pacification. Rather, they operated more subtly, chipping away the culture and ethos of the Bedouin tribes – particularly those practices of raiding and hospitality. This was not only about pacifying the desert, but pacifying the ethos of the Bedouin, making it less irruptive, dangerous and more capable of reconstruction as symbolic of (Trans-)Jordanianness.

To this end, Glubb employed an improvised form of disciplinary power. In particular, his methods revolved around what Foucault describes as the three means of correct training: hierarchical observation; normalising judgement and the examination (1991a: 170–194). Hierarchical observation operates by rendering subjects visible in order to transform them – 'to carry the effects of power right to them to make it possible to know them, to alter them' (ibid.: 172).

Thus, upon arriving in 1930, Glubb set out into the desert without his uniform to meet the most problematic tribes – the Huwaitat. In doing so, he relied upon their ethos of hospitality in order to befriend them, gaining trust and knowledge:

> I simply drove round all the camps of the Huwaitat, talking to them. Nomad hospitality had the immense advantage that one could stop at any tent and be immediately invited to dine and spend the night ... Amongst tented Arabs, it was possible to arrive as an uninvited guest in any family, to dine and sit up half the night talking in a relaxed atmosphere ... The situation was ideal for the application of my motto, *Love and trust and you will be loved and trusted.* (Glubb, 1983: 100)

In this way, Glubb won the confidence of the Huwaitat and persuaded them that they could not defy both the Saudi *Ikhwan* and the Transjordanian government (1948: 78). Glubb replicated his tactics from Iraq, where he had used Bedouin hospitality to get 'to know every tribe and village in my area so that, in the event of operations, I could lead aircraft to their target' (1983: 63). The 'love and trust' he employed was not benign; it was the initial step of pacification: building knowledge and ensuring visibility of the subject such that it could be disciplined. Thus, when in the March of 1931 a raiding party of Huwaitat crossed into Saudi Arabia, Glubb knew about it and sought to control it – 'I had so many friends among the Huwaitat that I always knew what was happening' (ibid.: 101). The hierarchical gaze operated, despite the desert context, even in the absence of Glubb through the surveillance of his 'friends'.

Glubb's nascent Desert Patrol had two tasks to achieve with the trust and knowledge they were building, both of which were about normalising the Bedouin population and Transjordanian state: stopping Huwaitat raids and organising them to defend themselves against *Ikhwan* raids. Having decided that compulsion was not working, Glubb withdrew British troops from the desert and sought the persuasion of a more patriarchal form of disciplinary power. His method was one of 'inviting the Huwaitat to run their own affairs', persuading them that they would be 'ruined, if not exterminated' if they did not 'stop raiding voluntarily, and organize defence measures'. The Huwaitat were therefore invited to join his Desert Patrol, policing, observing and normalising themselves instead of being 'policed by men from other tribes' (1948: 91–93). His logic was that of the autoimmune: they had to end their way of life in order to preserve it (Derrida, 2003).

Having helped organise a defensive formation at the border that would depend upon the Huwaitat's own cooperation and discipline (Glubb, 1948: 81–82, 94–95), the limits of such self-government became apparent. Here, the final stage of disciplinary training emerges: the 'examination', which combines the two previous components in a ritual testing and punishment (Foucault, 1991a: 184). The Huwaitat quickly broke both elements of their new disciplinary regime, abandoning the pickets required to defend the border (Glubb, 1948: 82–83) and engaging in return raiding (Glubb 1983: 101–102, 1948: 97–99).

Having failed their test, punishment would be that of normalising judgement: *corrective* rather than punitive (Foucault, 1991a: 179). Glubb dispensed with the ineffective legal threats of arresting and detaining raiders (1983: 102), opting instead for a patriarchal adaptation of the Bedouin's own raiding techniques. In failed examinations, the guilty Huwaitat had their camels stolen by Glubb and his desert patrol; when they came to apologise, a scene of hospitality played out where the Huwaitat were humiliated and subsequently forgiven:

> After looking grave for as long as our self-control lasted, we eventually winked at the solemn deputation, gave them back the camels, and told them to tell the Huwaitat not to be naughty, because next time we really would be angry. A large lunch of boiled mutton completed the proceedings. (Glubb, 1948: 99)

Through these disciplinary tactics of observation, normalised self-government and shaming punishment, Glubb began to pacify Transjordan's southern desert and borders. The gendered nature of this disciplinary pacification is clear. All the action, whether raiding, counter-raiding, punishment or hospitality is conducted by men and to men. Women appear in Glubb's account only as background to the Bedouin encampment. But crucially, the desert itself – that which is to be 'penetrated', mastered and brought under the normalising, hierarchical gaze, its 'virginity' thus 'violated' – is thoroughly feminised (Glubb, 1948: 85–87, 108).

From the Huwaitat, Glubb moved north to the more problematic tribes of the Jebel Druze mountains that border Syria. On the way, similar tactics of hospitality, persuasion and shaming were employed to prevent the border crossing and raiding of the Bani Sakhr (1948: 106–112). By this time, Glubb's adaptive disciplinary tactics included the production of 'delinquency' as a useful category of offender with which to normalise the rest of the population (see Foucault, 1991a: 257–292). Glubb is very clear that the aim was the destruction, indeed the criminalisation, of the Bedouin way of life:[11] 'In the past everybody raided, and raiding was a custom not a crime. From now onwards, we developed in a mild way a criminal class' (Glubb, 1948: 102). Again, the means of punishment were *corrective* and this allowed relatives to inform on their recalcitrant kin: they were forced to return all stolen camels, as well as one of their own as a 'fine to the Government'. Confessions were extracted by the method of examination – non-confession resulted in a harsher penalty (of six camels). 'This system abolished not only raiding but even ordinary stealing' (ibid.: 102).

By such disciplinary means, a 'firm but affectionate control over all the tribes' was established (Glubb, 1983: 105). However, the British and Hashemite guests had not fully understood the role that raiding played in Bedouin life. In 1932 a drought and subsequent famine struck the desert and revealed that raiding had acted as a way of redistributing material necessities. Rather than a 'pastime', raiding operated as a 'social-security system of which our ill-timed intervention had destroyed the balance' (Glubb, 1948: 168–169). The solution was, once again, corrective of the Bedouin's irruptive ethos: the nomadic tribes must be transformed into agriculturalists – at least semi-cultivators and stock breeders – whilst others were driven into joining the Desert Patrol and Arab Legion. The sedentarisation of the Bedouin required 'something of a minor revolution' as the 'fellah', or cultivator, was a term of abuse amongst Bedouin (ibid.: 169). Yet by 1939, Glubb claims that almost all nomadic Bedouin had become semi-sedentary. Such a revolution occurred 'in most cases' by persuasion; persistent delinquents, such as elements of the Bani Sakhr, had their camels and men rounded up at the point of a machine gun and, in 'punishment for their misconduct we sentenced them there and then to take up farming, and awarded them a large basin of land south-east of Amman' (ibid.: 170). Glubb's adaptive discipline was supplemented by the exercise of sovereign power, with the law literally made-up 'there and then'.[12]

The Desert Patrol's colonial adaptation of disciplinary power thereby abolished the centuries-old custom of raiding 'in a few months without inflicting a single casualty ... [or] putting a single tribesman in prison', while revolutionising/destroying their nomadic way of life. Meanwhile, the government, long regarded as the Bedouin's 'natural enemy', became their benign enslavers, enlisting thousands in the Arab Legion as Bedouin became 'the most loyal and patriotic citizens of Transjordan' (Glubb, 1948: 113). Thus we return to the two processes of pacification: destruction and reconstruction (Neocleous, 2011). The use and destruction of the Bedouin host's way of life was accompanied by their reconstruction as the symbols of a wider Jordanian space and ethos; 'the de-Bedouinization of the Bedouins as a precursor to the Bedouinization of Jordanian national identity' (Massad, 2001: 144).

## (RE)CONSTRUCTING A BEDOUIN ETHOS

While the pacification of the desert and Transjordanian borders allowed the space to firm up an inside and outside, evidenced by the beginning of formal diplomacy with Syria and Saudi Arabia in the 1930s (Glubb, 1948: 203–216), resistance to the guests' construction project was always present. Abdullah's attempts at exerting domestic control saw rising opposition from the 'Adwan, who felt he was favouring their historical rivals, the Bani Sakhr. In the autumn of 1923, this resentment built into armed rebellion with the 'Adwan marching on Amman and the Arab Legion repulsing them (Wilson, 1987: 77–78). A more serious resistance was faced by King Hussein from 1953 to 1957, as radical Arab nationalists, funded by Nasirist Egypt, coordinated a series of riots and protests culminating in an attempted coup led by one of Hussein's favourite officers (King Hussein, 1962: 101–183). Pacification of the desert was not enough; a specifically '(Trans-)Jordanian' ethos was needed to bind this space and promote a sense of belonging, one that distinguished Jordanians from the increasing number of Palestinian refugees (Alon, 2009).

King Hussein was especially influential in forming a new national narrative for postcolonial (Trans-)Jordan. Arriving back in Amman in 1952 after his father, King Tallal, had been deposed due to mental ill-health, Hussein reflected a desire to 'bring about more of a family spirit in my country' (King Hussein, 1962: 41). He 'planned from the very start to become the head of a family as much as the king of a country' (ibid.: 61), making explicit the patriarchal rule of Abdullah. This Hashemite metaphor of (Trans-)Jordan as a family home became engrained through its frequent repetition – 'they express[ed] their dominance in a patriarchal rhetoric brimming with kinship metaphors' (Shryock and Howell, 2001: 247). While sounding disingenuous to Western ears, it is an image of community 'that made immediate sense' to Hussein's subjects (ibid.: 247). Families, households and their heads continue to play an important role in Jordanian politics, which is still organised on tribal and familial lines (see Layne, 1994; Alon, 2009). Shryock and Howell (2001: 248) suggest that home and family have become so central to both the rhetoric and reality of Jordanian politics and society that they are 'best analyzed as a manifestation of "house politics," a mode of domination in which families (the royal one being only the most central and effective) serve as instruments and objects of power'. This 'house politics' is distinct from 'domopolitics' (Chapter 2) in being more literal – families/households are themselves influential political actors in (Trans-)Jordan and gain their status through their historical links of hospitality and patronage to the Hashemite family.

This patriarchal 'house politics' served another crucial function, allowing Hussein to massage the relation between (Trans-)Jordan and a wider Arab nation as that which exists between a family and tribe:

> When I think of my family, I think with pride of everyone in Jordan, who, standing by me as we faced the storms, inspired me in serving them. When I think of the tribe to which I belong, I look upon the whole Arab nation. My life is dedicated to a cause, just as the Hashemites have been throughout history; that cause is to be an Arab worthy of Arab trust and support … Remember that we are masters of our homes, and we were born free. (King Hussein, 1962: 99–100)

Hussein here negotiates continuity with his grandfather's passion – an Arab nation – without diminishing the 'way of life' he aims to produce as specific to (Trans-)Jordan (King Hussein, 1962: 99). Towards the end of this quotation he is speaking to the wider Arab nation – the tribe – whilst declaring his own sovereignty of, and belonging within, (Trans-)Jordan (the particular Arab 'home' of which he is 'master').

Massad (2001) offers a meticulous account of how an identity for this national 'home' has been built on the basis of a reconstructed version of the Bedouin – the 'Bedouinization' of Jordan. He places much of the responsibility for this on Glubb (Massad, 2001: 102–162), for whom the Bedouin were the most authentic of the Arabs as a martial race: 'the most typical surviving examples of that purely Arab way of life, which, amongst other Arab communities, has become … diluted by mixture with foreign influences' (Glubb, 1948: 9). Glubb was just such a foreign influence, a bad guest, central to breaking this way of life (as shown above) and then rebuilding a version as symbolic of national identity. His renovation was so powerful precisely because it began with targeting the most recalcitrant, irruptive 'Transjordanians': the Bedouin. He did so by recruiting them, almost exclusively, into the Desert Patrol and Arab Legion, a military force which became fundamental to securing (Trans-)Jordan from internal (tribal, Palestinian and alternative nationalisms) and external (Nasser's Egypt, Syria, Israel and Saudi Arabia) threats.

Many examples of the banal forms of the Bedouinization promoted by Glubb and Hussein are offered by Massad (2001), most of which can be found in Glubb's memoirs. Perhaps the most evocative example, however, is that of national dress. Glubb designed the Desert Patrol's uniform himself, claiming that it was the Bedouin's 'own natural clothing':

> white cotton trousers and a long white 'nightgown' or *thob*. Above this was a long khaki gown, a wide, red, woollen belt, a mass of ammunition belts and bandoliers, a revolver with a red lanyard and a silver dagger. The headgear was a red-and-white-checkered headcloth, which has since then (and from us) become a kind of Arab nationalist symbol. Previously, only white headcloths had been worn in Trans-Jordan or Palestine. (Glubb, 1983: 102–103)

Glubb notes that 'the effect was impressive' (1948: 103). The wider effects of such apparent banalities were indeed impressive. Before Glubb's fashion innovation, native Bedouin had favoured white or black-and-white *shmaghs*; Glubb's arbitrary selection of the red-and-white became a way of differentiating 'real' Transjordanians from Palestinian Jordanians who preferred the black-and-white *kuffiyeh* (Glubb, 1983: 103; Massad, 2001: 121). King Hussein would later wear the red-and-white *shmagh* at public events and on the cover of his autobiography (1962) to emphasise his Jordanianness, his belonging.

Massad, however, overstates the extent to which this Bedounization was colonially imposed solely from above (2001: 277–278). The nation- or ethos-building project was not forced upon the host tribes and notable families of (Trans-) Jordan entirely from without, via coercive means. Far from offering zero-sum conversion or resistance, Linda Layne (1994: 29) characterises the process as a 'series of cultural and political transactions' based in ongoing engagement. Yoav Alon (2009: 148) sees host–guest interactions as a 'negotiated relationship', starkly contrasting with the subjugation of tribal structures in Iraq, Palestine, Syria and Saudi Arabia. Glubb and Hussein's Bedouinization was successful because it worked with the Hashemite monarch's primary role of 'maintain[ing] the careful balance between the different tribes' and families that represent them (Alon, 2009: 154). This reconstructed 'Bedouinization' co-opted the host tribes through patriarchal persuasion and disciplinary transformation and was subsequently itself folded into the emerging 'house politics' that constituted the 'mode of domination' peculiar to (Trans-)Jordan (Shryock and Howell, 2001).

This Bedouinization required the post-hoc (re)construction of the Hashemites themselves in order to secure their place as masters of the 'house'. A central concern of King Hussein in his memoirs is thus to paint Abdullah as Bedouin: 'Abdullah's first love, it is true, was the Hejaz' (1962: 3), but he was 'a Bedouin at heart' (ibid.: 6). He longed for an 'intrepid Bedouin son' (ibid.: 19) and loved Bedouin coffee (ibid.: 21); 'a man of desert ways who had been brought up as a child among the Bedouin tribes' (ibid.: 17). This was an artful interpretation of an upbringing amongst the Ottoman elite of Mecca and Istanbul. But the fact that it is commonly accepted in the West and Middle East alike shows the success with which Abdullah's 'personal history became entwined with the creation of Transjordan and he came to be identified with and to identify himself with the tribal hierarchy of that country' (Wilson, 1987: 6; also Layne, 1994: 12).

King Hussein placed himself as a continuation of this tradition, living simply as is appropriate for one 'descended from the frugal life of the Bedouin', and whose only protocol of rule was that of the Bedouin – the 'three virtues – honor, courage and hospitality' (King Hussein, 1962: 63–65). The Hashemites' belonging, their right to be considered a host though they could more easily be viewed as guests, was secured as symbols and patriarchs of tribal Bedouin 'house politics'. Their hostness, however, could only be truly secured by casting out the rival guest – the British – and specifically Glubb as head of the Arab Legion. By 1956, Hussein was facing pressures from domestic nationalism which presented both British *and* Hashemites as guests who did not belong (Massad, 2001: 165–171). Hussein charts how his concern as the 'servant of Jordan' had turned to giving Jordanians increased responsibility, to 'give them pride in their country … and its future':

> Glubb, on the other hand, despite his love of Jordan and his loyalty to my country, was *essentially an outsider*, and his attitude did not fit at all into the picture I visualised. Yet, since the Arab Legion was the single strongest element in Jordan, he was, paradoxically, one of the most powerful single forces in our country. Consequently, to be blunt about it, he was serving as my commander-in-chief yet could not relinquish his loyalty to Britain. (1962: 131, emphasis added)

Glubb no longer fitted the reconstructed Bedouin ethos he helped create – he had no place within a 'house politics' with one patriarch and dominant family. Hussein admits that Glubb had been central to the formation of Jordan (1962: 150), but with Glubb leading the Arab Legion '[w]e were in the hands of foreigners' (ibid.: 137). He was thus dismissed and initially given two hours to leave the country (Glubb, 1957: 424). Whilst the Hashemites' own guest status would shortly be threatened by a failed coup, their belonging and sovereign mastery was obtained by ejecting their rival guest.

## SELLING AUTHENTIC *KARAM:* MARKETING JORDAN'S WELCOME

The manner in which guests and hosts negotiated the pacification, production and government of (Trans-)Jordan was part of a wider hospitality trend in the postcolonial becoming of the Middle East: the growth of tourism. In (Trans-)Jordan specifically, tourism became 'a tool for state building by helping the Hashemite monarchy define a national identity for the country and support its own claim to rule' (Hazbun, 2008: 79). Welcoming high-spending foreigners offered the opportunity to draw together many of the threads uncovered by the genealogy of (Trans-)Jordanian hospitality above: the ethos of the host; the pacified space of the homeland; the right of the Hashemite guest to rule; and the emerging nationalism based on a reconstructed Bedouin identity. This new ethos made of the Bedouin a 'fetishized commodity' (Massad, 2001: 120), such that symbols of Bedouin hospitality, particularly coffee-making paraphernalia, have become ubiquitous signs of welcome for outsiders in (Trans-)Jordan's public spaces (Shryock, 2004: 41).

Tourism has become central to the sustainability of (Trans-)Jordan. As a country with limited natural resources and agricultural capabilities, tourism was early identified as key to transforming its economy and reshaping its spatial environment (Hazbun, 2008: xi). It now constitutes the 'long term driver of economic growth in Jordan' (MoTA, 2011: 21), and is the sector contributing the most to GDP, rising from about 10 per cent in 2008 (USAID, 2008: 13) to 14 per cent by 2013 (USAID, 2013: 2). The role of commercial hospitality in sustaining and reproducing (Trans-)Jordan, as well as other postcolonial economies and spaces (see Hazbun, 2008; Carrigan, 2011), further demonstrates the inadequacy of IR's differentiated typology of state sovereignty. Sørensen (1997: 258) stipulates that a Westphalian sovereign state has a 'self-sustained' national economy, including the 'main sectors' (production, distribution and consumption) 'needed for its reproduction'. A state such as (Trans-)Jordan, dependent upon servicing guests' consumption of its products, is for ever condemned to a 'postcolonial', insubstantial sovereignty.

This reliance on tourism has opened (Trans-)Jordan to another key guest in the shape of USAID, the US government's aid agency, which has helped fund it since the end of the 1967 Arab–Israeli war (Hazbun, 2008: 84–131). In 2012, Jordan was the third largest beneficiary of USAID funds (behind Afghanistan and Pakistan), with nearly half a billion dollars in Obligated Program Funds (USAID, 2012). Much of this aid has been tied to investment in tourism, giving USAID unprecedented influence on the welcome Jordan offers. Thus, while in 2003 the

Jordanian Ministry of Tourism and Antiquities (MoTA) launched their 'National Tourism Strategy 2004–2010', this was developed with USAID support. In 2005, USAID designed and launched the Jordan Tourism Development Project, a three-year, $17 million scheme working with local and international actors to 'develop a dynamic, competitive tourism industry' (USAID, 2008: 14). A second version followed, running from 2008–2013 (USAID, 2013: 2). Examining these projects' final reports alongside the national strategy, we can see how tourists and USAID have had a profound effect on the space and ethos of (Trans-)Jordan, building on those produced by its previous guests – the (outcast) British and the (now-host) Hashemites. The latest guest's 'strategic vision' targets a new transformation – making Jordan a 'distinctive destination, offering year-round visitor experiences that will enrich the lives of Jordanians and their guests' (MoTA, 2011: 10; USAID, 2013: 10).

## DISCIPLINING AND UPGRADING HOSPITALITY

Jordanian space has been transformed in a variety of ways by this influx of private and public investment. Much investment was concentrated on the significant existing 'sites', especially Petra, the Amman Citadel, Jarash, Wadi Rum and Aqaba. But work has also been done to develop Madaba, Aljoun and Salt into 'sites' worth visiting (USAID, 2013: 28–67). This transformation involved the displacement of local populations (especially the tribes native to Petra) and disruption to local communities caused by building tourist facilities, such as bus parking and rest areas (Al Haija, 2011). Clearer tourist signage, landscaping, infrastructure, visitor centres, museums and tourist trails have been developed to make them more accessible for guests. USAID helped secure protected status for Wadi Rum as a UNESCO World Heritage mixed site, restricting its uses for guests and hosts alike (USAID, 2013: 53). When one adds to this the appearance of ostentatious symbols of welcome – such as the giant 'traditional' Bedouin coffee pots Shryock (2004) identifies in public spaces – we see how the changing space of (Trans-)Jordan, similar to parts of the global city (Chapter 3), works to accommodate the whims and requirements of guests over those of hosts. Like USAID's description of the Amman Citadel, the whole of (Trans-)Jordan is becoming 'an astounding open-air museum' (2013: 44).

Perhaps the more profound transformation, however, has been aimed at disciplining the (Trans-)Jordanian ethos of hospitality. In USAID and MoTA reports, the 'natural hospitality' of Jordan's people are seen as key to its appeal as a 'destination' (MoTA, 2011: 8; USAID, 2008: 13, 2013: 121). The 'spirit of its welcome' is central to its 'brand development' and 'integrity' (MoTA, 2011: 39). 'Jordan's unique selling point is "Jordanians"; the Arab hospitality that we are not used to in the US' (Peter Greenberg in USAID, 2013: 120). From an ethos of *karam*, a way of being and dwelling in relation to others as guests and a complex negotiation of sovereignty, hospitality has become a marketable asset, a unique selling point (USP). In trumpeting the success of its projects, USAID cites Um Ahmad, a single-mother-of-three who has made her home into a bed and breakfast on the Aljoun tourism trail (ibid.: 63). USAID helped her develop her business, offering 'a touch of warm authentic Jordanian hospitality' to 'paying guests'. Of course, Bedouin hospitality, which has become 'Jordanian hospitality', can *never* be paid for; this is business, not hospitality, and is seen as 'dirty work, akin to prostitution' (Shryock, 2004: 41). While Um Ahmad's endeavours are admirable, portraying them as an offer of 'authentic' hospitality that can be bought is deeply problematic.

Despite the centrality of this 'natural' ethos of hospitality to its brand, USAID has concentrated on correcting, training, and thus once again *disciplining* (Trans-)Jordanian hospitality. The aim of USAID's hierarchical observation and normalising gaze is to bring this hospitality up to 'international' standards. A key 'Action' point of the National Tourism Strategy is to 'build hospitality skills across all sectors of the tourism industry' (MoTA, 2011: 70). To this end, vocational training programmes and centres were introduced or revamped, including internships and leaving certificates (USAID, 2008: 25–30), with educators and employers collaborating to initiate or improve hospitality education in secondary schools, colleges, public and private universities (MoTA, 2011: 66). A great push was made to target the recruitment of women into the hospitality sector at all levels (MoTA, 2011: 15, 30, 63–65; USAID, 2008: 10–27, 2013: 2–3, 73–83), changing the traditional gendering of 'Jordanian' hospitality. One area of particular concern was the Bedouin camp sites at Wadi Rum, where tribesmen experienced special training in hospitality (USAID, 2008: 20, 39). American experts thus taught Bedouins how to practise authentic hospitality.

The disciplinary aim of 'benchmarking' and lifting Jordan's touristic 'product' up to 'international standards' (MoTA, 2011: 14–17, 49; USAID, 2008: 13–15, 2013: 15) is precisely *normalisation* and homogenisation (Foucault, 1991a: 184). Yet the stress throughout is always on developing the 'product' to emphasise its distinctiveness and *authenticity*, whether this be as cultural experience (MoTA, 2011: 55), food and customs (ibid.: 57), parades and festivals (USAID, 2013: 3) or 'Jordanian hospitality' itself (ibid.: 61). This produces a fascinating incongruity between drives for homogenisation and distinctiveness, falsity and authenticity. The final report on USAID's first project reflects the great strides made in 'upgrading' Bedouin campsites in Wadi Rum using 'minimum standards' to 'meet the expectations of international tourists' (2008: 37). The 'examination' element of disciplinary tactics, where hierarchical observation and normalisation meet in a ritual testing, occurs in the inspection of campsites. Twelve of 37 Bedouin sites passed their inspection, with 22 initiating the procedure (ibid.: 20). Without such upgrading these camps do not receive a licence to operate (ibid.: 39). A simple system of punishment and reward thus completes the disciplinary procedure. The minimum requirements for a licence included standards of safety, hygiene, comfort, environmental measures and 'authenticity' (ibid.: 20). The fact that such minimum standards undermine the authenticity of the Bedouin camp site is ignored. The photograph of an 'upgraded' camp (Figure 4.1) shows what looks like a prefabricated building covered in traditional black goatskin, atop a brick foundation. You can now experience the 'authentic' hospitality of the nomadic Bedouin in a tent with immovable brick walls.

**Figure 4.1** 'Authentic' Bedouin camp site at Wadi Rum (USAID, 2008: 38)

A similar dance around authenticity is performed regarding 'cultural' souvenirs; these require an 'upgrade' in order to offer 'high quality authentic designs' (MoTA, 2011: 50). Handicrafts are an 'important part of the tourist experience', but 'the quality of handicrafts is not always up to international standards' (USAID, 2013: 68). Fortunately, the American guest has developed a regulating disciplinary mechanism, the National Handcraft Strategy, which trains artisans and 'encourages' them to produce new designs while 'creating a national identity through handcrafts' (ibid.: 70). Once again, the Bedouin are particular targets of this regulation and identity construction. Bedouin crafts are especially prized, though implicitly criticised for being sub-standard and requiring USAID improvements:

> The Bedouin of Wadi Rum have a rich culture and heritage, much of which is manifested in traditional handmade products and handicrafts. To improve the tourism potential of this aspect of Bedouin life, USAID provided technical assistance to CBOs [community-based organisations] in handicraft production to improve and diversify designs and ensure availability of

> equipment to produce handicrafts. As a result, artisans at Al-Diesseh, Productive Village, and Burda cooperatives made significant strides in handicraft design for leather, woven goods, ceramics, jewellery, soaps, and traditional items. (USAID, 2008: 20)

The patronising and paradoxical rendering of this rich culture making 'significant strides' in producing their own authentically 'traditional items' is underlined in USAID's final report (2013: 56), with a training manual now available to help develop 'new' but 'good quality authentic Jordanian handcrafts' (ibid.: 71). As Shryock notes of one women's cooperative venture emerging from these projects, none of these products are 'remotely traditional, but the people who make them are easily portrayed as traditional (if not altogether backward), and this certainly adds value' (Shryock, 2004: 46).

The USAID guest is engaged in an elaborate and expensive project which seeks to make all aspects of (Trans-)Jordan's hospitality visible to its normalising gaze, reconstructing an 'authentic' ethos that is homogeneous enough to be sold to international tourists. This pacifying, disciplinary effort is layered on top of the projects of guests past (Britain) and present (Hashemite). Whether this be through teaching 'naturally' hospitable people how to practise (and regulate) their hospitality or training them to produce (and completely change) their cultural products, the aim is a more acceptable version of themselves, packaged for their guests.

## RESISTING NORMALISATION

Resistance to the disciplinary tactics of this reconstructed ethos can take many forms (see Al Haija, 2011), but the most prominent has simply been non-engagement, born of a cultural antipathy toward the hospitality industry. There has historically been a large gap between the tourism jobs on offer and the skilled workers to fill them – 'The main reason for this is the existence of … a "culture of shame" towards working in the tourism and hospitality industry' (USAID, 2008: 13, 2013: 80). The industry has a 'poor image', due to a 'lack of understanding' according to the government (MoTA, 2011: 65). This has resulted in families (especially men) discouraging their children and women from working in hospitality. Amongst the Balgawi Bedouin, families are particularly concerned that females are being used as cheap labour which will bring them into inappropriate contact with male labourers, and as 'bait' to entice both tourists and foreign aid (Shryock, 2004: 46). The women that do work in Foundations and Cooperatives often withdraw if their privacy is threatened or the work too demeaning.

The governmental tactic employed to alter this conservative ethos was a campaign of public re-education (USAID, 2008: 45; MoTA, 2011: 79). The National Tourism Awareness Strategy was a nationwide suite of media campaigns, awareness workshops, poster competitions and training seminars entitled 'Tourism is

Everyone's Business'. It initially targeted 'groups that played influential roles in changing perceptions of tourism', before focusing on 'Jordanians at large'. USAID concluded that, by 2008, 'more than two million Jordanians have become more aware of the importance of tourism' (2008: 45–46). A renewed campaign followed from 2009–13 (USAID, 2013: 101–102). The follow-up national survey determined the tactics' success, with over 90 per cent of parents agreeing that jobs in tourism were not 'shameful' (ibid.: 104), while many more Jordanians were training for or entering tourism professions, especially women (ibid.: 83). The true accomplishments of these governmental tactics in shifting the ethos of 'Jordanians' is hard to judge, especially amongst Bedouin. Shryock found that attitudes amongst Balagwi communities were changing, but in more subtle ways. While ostentatious displays of commercial hospitality were accommodated as 'business', this was not allowed to impinge upon what they saw as the more authentic realm of hospitality – the family home. Here, gender roles, privacy and sovereign mastery were carefully guarded. The role of females in the 'business' side was negotiated rather than accepted, such that their becoming like the products they sell, on display for tourist consumption, was still resisted and rejected (Shryock, 2004: 46–48).

As the ethos and space of (Trans-)Jordan continues to change, moving beyond itself through the influence of guests that are both valued and suspect, more violent forms of resistance remain a possibility. Even if subdued, there are plenty of cracks and fissures in the management of this homogenised Jordanian hospitality for those who look beneath its shiny surface. To take a simple example, one of the 'success stories' offered by USAID is that of Sahar Abu Nassar and her shop, 'Saher Al-Sharq' (2008: 17). Her products are 'inspired by traditional embroidery' using 'intricate patterns and colourful designs' sewn into a variety of items. With help from USAID, she has built a business in Madaba, employing a network of over 50 women. USAID does not specify whose 'tradition' these products are inspired by, but a visit to the Seher Al-Sharq website,[13] invites us to discover 'the rich history and heritage of Palestinian embroidery'. Viewing 'Sahar's Story',[14] we find that the tradition preserved by Sahar is that of Palestinian refugees who fled to Jordan, and the 'rich heritage of places like Jerusalem, Bethlehem, Jaffa and Ramallah' – none of which is in contemporary Jordan. Even the authentic Jordanian traditions USAID is selling tourists are those of the self-identified outsider, the guest of (Trans-)Jordan which, in various guises (British, Hashemite, American, Palestinian) and throughout its history, has sought to design, construct and transform this space and its ethos.

## CONCLUSION

Guests are once again transforming the space of Jordan. According to the Migration Policy Centre, 622,000 Syrian refugees had sought hospitality in Jordan by early 2015 (Achilli, 2015), though estimates vary up to 1.3 million

(BBC, 2015a). Eighty-four per cent are living in cities, mutating the urban environment, with the remaining 16 per cent based in five official camps, the largest of which is Zaatari near the Syrian border (ibid.: 5). Nowhere is the spatial transformation clearer than in Zaatari, set up in the summer of 2012 to receive Syrians. At one time reported to be the fourth largest city in Jordan (Weston, 2015), Zaatari remains home to over 80,000 refugees. The *New York Times* has reported it as a 'do-it-yourself metropolis', with 'neighbourhoods, gentrification, a growing economy and ... something approaching normalcy' (Kimmelman, 2014).

Despite growing concerns that the government is restricting hospitality, closing its borders to new arrivals (Achilli, 2015), (Trans-)Jordan continues to be praised for its near-unconditional hospitality. Such acclaim normally places this hospitality within a narrative of the 'natural' hospitality of Arabs/Jordanians and the history of welcoming Palestinian and Iraqi refugees. But the genealogy of Jordan's hospitality is far more complex, contested and *productive* than this simplistic narrative suggests. This chapter has offered a different line of descent, focusing on three episodes. The first looked at the encounter after World War I between the British mandatory administration, the Hashemites and the native tribes of this ostensible 'no man's land'. What emerged was an alternative history of clashing *ethea*, of 'good hosts' and 'bad guests'. The second episode examined British and Hashemite guests' pacification of the desert and cooptation of a reconstructed 'Bedouin' ethos to define Jordanianness through a negotiated, patriarchal 'house politics'. The final episode explored the welcoming of the USAID guest to once again reconstruct and upgrade the space and ethos of (Trans-)Jordan in order to market its hospitality to high-paying tourists, upon whom postcolonial states often rely.

This genealogy demonstrates how a problematically post/pre/non-sovereign postcolonial space of (Trans-)Jordan has been formed out of the near-unconditional hospitality of its subaltern host and the imperial, patriarchal discipline of its guests. This disciplinary gaze continues today, with USAID and the government normalising and homogenising its welcoming of tourists. I am not claiming that this constitutes a loss of 'authenticity', the forfeiture of a 'truly' native art of hospitality. Rather, I am arguing that Jordan's near-unconditional hospitality to Syrian refugees must be understood within the power relationships between hosts and guests which *produced* the contemporary 'Jordan' that now receives such praise. Furthermore, this hospitality is again being enabled by outsiders. Primary amongst them is the EU and its politics of protection which funds postcolonial states such as Jordan to contain refugees far from Europe's borders. Thus, when King Abdullah II (2015) expresses his thanks to 'all of you who are helping us to uphold this global responsibility' of hospitality, he is speaking before the European Parliament. The final chapter therefore turns to the post-sovereign space of EUrope and its (auto)immunising practices of hospitality.

## NOTES

1 Walters (2012: 114–115) notes that Foucault was not firm or consistent in his understanding of 'genealogy' as a method. Three 'styles' of genealogical inquiry – as descent; as re-serialization and counter-memory; and retrieval of subjugated knowledges and struggles (2012: 112) – are discernible, though they are not mutually exclusive and often involve improvisation (2012: 117). I do not consciously employ one style over another, but what follows is probably closer to re-serialization and counter-memory.
2 Amir of Transjordan under the British Mandate (1921–46); King of Jordan until his assassination in 1951.
3 King of Jordan from 1952 till his death in 1999.
4 Sir John Bagot Glubb entered Jordan in 1930 as an officer in the Arab Legion, setting up his 'Desert Patrol' in 1931 and becoming commander of the Arab Legion from 1939 until he was forced to leave Jordan in 1956.
5 Herbert Samuel was the first High Commissioner of Palestine under the British mandate (1920–25).
6 The label '(Trans-)Jordan' signifies two things. On the one hand, it refers to the shifting identity and borders of this space, which have changed three times. Its formal name changed from the Emirate of Trans-Jordan (1921–46) to the Hashemite Kingdom of Transjordan (1946–49) to the Hashemite Kingdom of Jordan (1949–). Its borders shifted first in 1925, incorporating the regions of Aqaba and Ma'an from the Hijaz. Second, in 1950, formally annexing the West Bank during the Arab–Israeli war of 1948. Finally, the latter was relinquished to Israel following the Six-Day War of 1967. On the other hand, retaining the 'Trans' label helps to underline that this fluid space does not fit into Eurocentric ideal-types of Westphalian states and 'postcolonial' or 'quasi' states. Instead, 'trans' as a prefix refers to its status 'beyond', or 'across' such ideal-types, always in process of 'completely changing'.
7 King Hussein (1978: vii–viii) offers one description of 'natural Syria', but a more precise mapping of the Arab homeland is offered by Abdullah I: 'Syria and Iraq are respectively the western and the eastern frontiers of the Arab lands. Syria lies on the Mediterranean coast and is bounded on the north by Turkey. Iraq, on the eastern frontier, is bounded on the east by Iran and on the south by the Persian Gulf ... form[ing] an important political and geographic unity. To the south of them are the deserts of Najd and the Hijaz ...' (King Abdallah, 1978: 27).
8 This conceptualisation has been stripped of its Weberian trappings and given an overtly teleological, interventionist and Imperialist rendering in the work of the influential British and EU-based diplomat Robert Cooper. He describes the three types as 'pre-modern', 'modern' and 'postmodern' (see Cooper, 2003: 16–54).
9 Cooper's replication of Sørensen's framework (see previous footnote) was particularly influential. He became special adviser to Tony Blair, Head of the Cabinet Office's Defence and Overseas Secretariat, the UK's Special Representative in Afghanistan in 2001–2 and later an EU diplomat. During this time he called for 'Defensive Imperialism' against the failure of pre-modern states (Cooper, 2002: 19).

10 Though I am substantiating my claims using the original memoirs of the British and Hashemite guests, I am primarily allowing an American anthropologist (Shryock) to speak for the hosts. This is deeply problematic, though it does illustrate the more subaltern nature of the *host* in this context, closer to the silent, spectral *guest* in Chapter 3.

11 Massad offers a meticulous account of the many ways in which Bedouin life was later criminalised from 1958 to 1976 (2001: 58). These tactics built on Glubb's earlier work.

12 Glubb's sovereign authority over the Bedouin was authorised through the Law of Supervising the Bedouins (passed just before his arrival in Transjordan, in 1929). This 'effectively put all power in relation to the Bedouin population in the hands of the head of the army, or his deputy (in this case Glubb), thus relegating all Transjordanian Bedouins to living under martial law' (Massad, 2001: 117).

13 www.sehralsharq.com/home-.html (accessed 15 July 2015).

14 www.sehralsharq.com/sahars-story.html (accessed 15 July 2015).

# 5

# (Auto)Immunising Hospitality: EUrope

EUrope's hospitality is in crisis.[1] Over 600,000 refugees arrived in Europe via the Mediterranean Sea alone in 2015, with over 3,000 dead or missing.[2] Nearly half of these people are thought to be from Syria, with many fleeing from Afghanistan, Iraq, Eritrea and elsewhere (IOM, 2015). The scale of arrivals, and EUrope's inability to welcome them in a coordinated manner, have led to characterisations of a 'migrant crisis' (BBC, 2015b), a 'refugee crisis' (*Guardian*, 2015a) or a 'border crisis' (Vaughan-Williams, 2015) for the continent (Europe) and its institutions (the EU). EUrope's response has been marked by in-fighting and the closing of national borders, all of which is jeopardising long-term EUropean solidarity. Jean-Claude Juncker, President of the European Commission, has been withering in his criticisms of member states' hostility to refugees, characterising it as un-European (Juncker, 2015a, 2015b).

Perhaps surprisingly given these events, the metaphor of Europe as a home with the necessity of welcoming the outside world is common in EUropean discourse. The EU is also used to being identified with, and speaking self-confidently of, its norms, values and ethics (Manners, 2002, 2008; Lucarelli and Manners, 2006; Williams, 2010), which are expressed through its hospitality (Bulley, 2009). Over recent decades EUrope has developed a highly conditional and bureaucratised practice of welcoming particular types of pre-identified subjects: states that have a 'European perspective' or 'vocation' (Rehn, 2005e, 2006d, 2009b; European Commission, 2008a; Füle, 2010b; Hahn, 2015). Such hospitality is considered not only the most successful of EUrope's foreign and security policies, but also the very epitome of its ethical vocation – an expression of its ethos as a community of values. But individuals and communities not organised as recognisable states (such as refugees and migrants), or having no 'European perspective' (such as Morocco and Ukraine), are rarely welcome. The multifaceted nature of EUrope's hospitality is thus starkly revealed in the current crisis, especially when read in conjunction with EUrope's continued enlargement.

My central argument in this chapter is that EUrope's hospitality operates as an immune system which both produces and *protects* the space and ethos of a EUropean communal home against the dangers of instability coming from outside. Immunisation, according to Roberto Esposito, is always a 'protective response in the face of risk' and is intimately related to community through the shared etymology of *immunitas* and *communitas* in *munus* – the law of reciprocal exchange (2011: 1–7). Because of its immunising logic, EUrope's hospitality works differently to the other spaces examined in this book, operating in liminal spaces neither fully inside nor outside its 'home'. However, because EUrope's ethos is constituted by antimony and ambivalence, the hospitality which guards it ends up also attacking it, revealing an irreducible *auto*immunity. Autoimmunity describes the 'strange illogical logic by which a living being can spontaneously destroy, in an autonomous fashion, the very thing within it that is supposed to protect it against the other, to immunize it against the aggressive intrusion of the other' (Derrida, 2005b: 123). Thus what we see in EUrope's autoimmunising hospitality is that the home both protects and attacks itself; indeed, that it attacks itself through that very protection.

I explore the tensions of this hospitality through EUrope's interpretation of itself and the practices of welcoming and excluding expressed in treaties, policies, Commission and European Council Presidency communications and Commissioner speeches. The chapter proceeds by first considering EUrope as a post-sovereign, shifting space, whose indefinite ethos and values necessitate an open, if immunising, welcome to the outside world. The second section outlines how EUrope's hospitality has played out through its 'Enlargement' policy, welcoming states that can demonstrate their belonging within the EUropean home. EUrope makes use of the 'road' towards its door as a way of immunising itself against difference, transforming the other into the self *before* entry is permitted. Though seemingly generous, the power relations of this heavily conditioned welcome illustrate how it aims to inoculate EUrope from the threat of instability. The third section focuses on EUrope's immigration and asylum policies which create EUrope as a 'space of protection' via a form of humanitarian government. The vast majority of this protection is, however, offered outside the home. Protection is provided in spaces that are *becoming*-EUropean, through Regional Protection Programmes that raise standards in surrounding countries while guarding EUrope from being overrun by potentially threatening refugees. The final section concentrates on how these immunising practices are being resisted *from within*. The ethos of EUrope turns on itself, its contradictory logics deepening the current crisis, making it one of autoimmunity.

# EUROPE AS SPACE, HOST AND ETHOS

Drawing on Europe's origin myths, Zygmunt Bauman (2004: 2) discerns a 'common thread': 'Europe is a mission – something to be made, created, built.' This idea of Europe as unfinished reveals the instability and ambiguity at its heart, that there is no blueprint or road map, no clear sense of what Europe is or will become. Such indiscernibility has prompted a plethora of analyses of the idea of Europe (Derrida, 1992b; Delanty, 1995; Heffernan, 2000; Christiansen et al., 2001; Amin, 2004; Zielonka, 2006; Habermas, 2009; Steiner, 2015). When we refer to 'Europe' it is not clear whether we are talking about a geographical territory, an institutional actor, an idea, concept or essence (Bauman, 2004: 6). I have argued in this book that hospitality is a spatial relation with affective dimensions, demanding a space that can be both closed and opened by a host, and an ethos which helps define belonging and non-belonging within this space. The two elements of space and affective belonging are closely intertwined and deeply vexed in the production of EUrope.

## EUROPEAN SPACE

The European integration project has always had the 'reconfiguration of political space' at its heart (Bialasiewicz et al., 2005: 333). The problem is the frequency with which this space is reconfigured and the many forms it takes. While EUrope's understandings of itself as a 'home', 'family' and 'community' are relatively consistent, with the area to its south and east officially termed its 'neighbourhood', the boundaries of this space are not finally determined. When the EU's representatives refer to 'Europe', they could be talking about the EU's 28 member states; those 19 states that make up the Eurozone; those 22 EU and 4 non-EU states that make up the 'borderless' Schengen Area; or even the 32 states involved in coordinating spatial planning in the EU through the European Spatial Planning Observation Network (ESPON).[3]

The first, seemingly simple option is not clear-cut. The outer boundaries of the EU keep changing as new members join, Croatia most recently in 2013, while the Treaty of Lisbon opens up the possibility of a member's withdrawal.[4] Meanwhile, the EU's regional policy, building 'macro-regions' such as the 'Mediterranean' which encompass parts of the EU *and* its neighbourhood, has blurred 'hard' boundaries between inside and outside (Jones, 2011; Bialasiewicz et al., 2012). EU borders have not only blurred, they have been exported, off-shored and projected way beyond the threshold of 'Europe'. Over the last decade, the EU's management of migration, particularly through the operations of its border agency, Frontex, has been operating in Africa and the Middle East to deter and confine would-be migrants (Bialasiewicz, 2012; Vaughan-Williams, 2015). This is why I refer to 'EUrope', rather than the 'EU' or 'Europe', as this better captures the ambivalent relation between the geography of the space and its fluctuating institutional configurations (Clark and Jones, 2008; Bialasiewicz, 2011; Bialasiewicz et al., 2012; Vaughan-Williams, 2015).

The possible cartographic representations of these reconfigurations of EUropean space are legion. Perhaps most interesting is ESPON's vision for the territory in 2050 (see Figure 5.1), described as an 'open and polycentric Europe', 'cosmopolitan', 'connected' and welcoming to 'the rest of the world' (ESPON, 2013: 12). This representation includes no borders in order to indicate that current constraints on 'territorial development and government have disappeared, and the European Union remains open to internal and external enlargement processes' (ibid.: 20). EUrope therefore spills out, predominantly eastward into Russia and Central Asia, but also into North Africa and the Atlantic. Nonetheless, the various travel, energy and business relations represented by the different coloured lines are much denser within the current EUropean home. This map obviously misses out crucial elements of EUropean space and yet, as with the 'Key Diagram' of London (Figure 3.1), it shows that EUrope, like the global city, leaks at its edges and contains no vision of its final determination. Such territorial elusiveness produced political and conceptual confusion in the negotiations of the Constitutional and Lisbon Treaties, caused by 'the lack of a clear and direct correspondence between sovereignty, identity and territory. Europe's "undefinability" ... was interpreted as *inexistence* or, at best, lack of *purposeful existence*' (Bialasiewicz, 2008: 72).

EUrope negotiates this impasse precisely through the entanglement of space and ethos, territory and values. It has consistently over the last 15 years interpreted itself to the world as a space of *values*. These values define EUrope's way of being and belonging, its ethos. This was very much evident from 1999 to 2004, with EUrope united on the basis of shared 'ethical and political values' (Prodi, 2000a; see Bulley, 2009). But it has been sustained more recently with Olli Rehn, Commissioner for Enlargement from 2004–2010, and his successor Štefan Füle:

> I am often asked where Europe's borders lie. My answer is that the map of Europe is defined in the mind, not just on the ground. Geography sets the frame, but fundamentally it is values that set the borders of Europe. Enlargement is a matter of extending the zone of European values. (Rehn, 2005a)

Thus, EUrope is 'above all a community built on a set of principles and a set of values' (Solana, 2000a), a 'common home of shared values' (Füle, 2010d).

**Figure 5.1** An 'open and polycentric' EUrope in 2050 (see ESPON, 2013: 21) © ESPON 2013, Origin of information: ET2050 ESPON project, ESPON 2013 Programme with the European Regional Development Fund

These shared values which define its spatiality are understood as an ethos, a way of being in relation to self and others; they express 'a particular European way of life' (Rehn, 2009d), governing its 'transformative' relation to its neighbourhood (Rehn, 2009e) and 'all our partners' (Füle, 2010d). And crucially, this ethos is explicitly hospitable and welcoming; after all, the 'European Union has never been about building walls but about eliminating dividing lines through values and principles' (Füle, 2013a).

## A EUROPEAN ETHOS

So what are the values that define the home, community and ethos of EUrope, and which it seeks to protect through practices of hospitality? They are frequently listed by EUrope's institutional representatives in their speeches. These lists can be restrictive, including only democracy and the rule of law, but can also capture solidarity, peace, tolerance, human rights, fundamental freedoms, the protection of minorities, justice and equality. These are the principles on which the EU is 'founded', according to Javier Solana (2001, 2009a) – the first High Representative for the Common Foreign and Security Policy – and which it seeks to 'project' in its foreign relations (Solana, 2001). For evidence of this value-laden ethos, Rehn (2005a) suggests that we look to the EU's treaties.

However, even here the picture is as fuzzy as EUrope's territorial borders. The Treaty of Paris (1951), which established the European Coal and Steel Community, the earliest forerunner of the EU, offers no mention of 'founding values', only a reference to 'peace' and 'solidarity' in the preamble. Likewise, the Treaty of Rome (1957), establishing the European Economic Community (EEC), contains no explicit mention of 'values', though its preamble confirms 'the solidarity which binds Europe' and resolves to 'preserve and strengthen peace and liberty'. Democracy, the rule of law, human rights, equality and social justice only appear in the preamble to the Single European Act of 1987, though not as 'founding' values. The preamble of the Maastricht Treaty (1992), or Treaty on European Union (TEU), which properly established the EU, declares an 'attachment to the principles of liberty, democracy and respect for human rights and fundamental freedoms and of the rule of law' and a *desire* to 'deepen the solidarity between their peoples'. These principles would subsequently become a condition of being considered for membership of the EU, when the Copenhagen European Council set out its criteria. Alongside economic measures, potential candidates must have 'achieved stability of institutions guaranteeing democracy, the rule of law, human rights and respect for and protection of minorities' (European Council, 1993: 7.A.iii).

It was the Treaty of Amsterdam's (1997) amendment of the TEU which, for the first time, raised this into the body of the text, becoming Article 6(1), and claiming that 'The Union is founded on the principles' outlined in the 1992 preamble. As Andrew Williams (2010: 7) notes, in claiming such a foundational role for values, the Treaty is making 'an assertion that was not necessarily self-evident'. The Lisbon Treaty changed these 'principles' to 'values', becoming Article 2 of the TEU:

> The Union is founded on the values of respect for human dignity, freedom, democracy, equality, the rule of law and respect for human rights, including the rights of persons belonging to minorities. These values are common to the Member States in a society in which pluralism, non-discrimination, tolerance, justice, solidarity and equality between women and men prevail.

Here we have a clear expression of EUropean 'values', and this is supplemented by Article 3(1) TEU, which specifies the aims of the EU as promoting 'peace, its values and the well-being of its people'. Further, these values will, as Derrida notes of an ethos, define its relationship to itself and others – 'In its relations with the outside world, the Union shall uphold and promote its values and interests and contribute to the protection of its citizens' (Article 3(5), TEU). Thus, the first post-Lisbon High Representative for Foreign Affairs and Security Policy, Catherine Ashton, claimed that EUrope's external relations were 'built' on EUrope's 'basic values' – 'They are a silver thread running through all that we do' (Ashton, 2010b).

Despite such grand proclamations, as evidence of a foundational EUropean ethos the treaties are tenuous: the claim that the Union is 'founded' on a set of values only declared in 1997, 46 years after the Treaty of Paris, bears little scrutiny. Furthermore, none of these values are given any definition or explanation as to *what* they mean, *why* they in particular have been chosen, and in what *way* they are significant. No guidance is offered for when they necessarily clash – for instance, when freedom clashes with democracy because people vote to restrict freedom; when internal solidarity clashes with respect for the international rule of law, EUrope's contribution to a 'wider solidarity amongst peoples', or the protection of outsiders' human rights and dignity. Ultimately, this listing of values demonstrates that, like democracy (Derrida, 2005b: 8), EUrope's ethos is a 'meaning in waiting, still empty or vacant'. While certainly expressing *something*, at its centre is a 'semantic abyss that ... opens onto all kinds of autoimmune ambivalences and antinomies' (ibid.: 72). Williams is scathing in his criticism of the process, but also notes some hope:

> Identifying such a plethora of constitutional principles and values mixed with policy statements is a particularly inept way to construct, or even simply represent, a meaningful philosophical framework for the EU. There is little by way of definition here that might counter the uncertainty ... Nonetheless, with the Lisbon Treaty provisions coming into force, there is a clear and concerted attempt to enshrine constitutionally a notion of the 'good' for Europe that is sought through the EU. (Williams, 2010: 8)

While there is no evidence of a public ethos shared by citizens and civil society throughout EUropean territory, these values do express an *institutional* ethos, the collective character, values and disposition of EUropean *institutions* (Williams, 2010: 10–13).

This is a useful distinction. These values, indefinite and unstable though they are, produce the sense of affective belonging necessary to constitute EUrope as an institutional home where some belong and others do not. Furthermore, it underlines the use of the term 'EUrope' as the relevant space of hospitality. It is not the space and ethos of 'Europe' which is excavated in this chapter, but its institutional interpretation and representation as EUrope, defined by an imprecise and uncertain set of values through which it constructs its own identity and history. Michael Heffernan (2000: 6) argues that 'Europe' is best interpreted as a 'contested geographical discourse; as a series of invented geographies which have changed over time and across space'. In this sense, a EUropean space of values is just one possible invention of Europe more broadly. This geography is best expressed by former Commissioner Vladimir Špidla: 'Europe ends' where its values 'are not shared' (quoted in Williams, 2010: 3).

## THE HOSPITABLE PRODUCTION OF EUROPE

It is notable that this invented geography is institutionalised in the late 1990s. With the fall of Communism and the emergence of Central and Eastern European (CEE) countries from Soviet domination alongside rising fears about illegal immigration, the 1990s was a time when EUrope confronted different external dangers. These generated renewed attempts to both define 'Europe' and *protect* it. When a group of intellectuals organised by the Institut für die Wissenschaften vom Menschen in Vienna were asked by the EU to reflect on European identity in 2002, their final report exposed the productive role of this encounter with difference.

> What is European culture? What is Europe? These are questions that must be constantly posed anew. So long as Europe is of the present, and not simply the past, they can never be conclusively answered. Europe's identity is something that must be negotiated by its peoples and institutions … *Europe and its cultural identity thus depend on a constant confrontation with the new, the different, the foreign. Hence the question of European identity will be answered in part by its immigration laws, and in part by the negotiated accession terms of new members.* Neither of these – either the immigration laws or the terms of accession – can be determined a priori on the basis of fixed, static definitions, such as a catalogue of 'European values'. (Biedenkopf et al., 2004: 8–9, emphasis added)

Despite the reflection group's concerns, a catalogue of values is precisely what EUrope returned to in defining its ethos and identity. But this remains the result of a confrontation with difference; it forms a guide to how that which does not 'belong' is to be welcomed inside or excluded: policies on enlargement and immigration in particular. The ethical space of EUrope is thus produced through its practices of hospitality and hostility. Crucially, however, this negotiation of what/how EUrope welcomes not only produces EUrope, but also guards it. Its values act as an immune system, 'the strategies it employs to protect itself' (Hagglund, 2008: 13). And such strategies have emerged in relation to the possible threat from outside.

This is demonstrated if we look at the policies in question. Cooperation on migration and asylum began with the Maastricht Treaty in 1992, under the

Justice and Home Affairs pillar. Post-sovereign supranational decision making in this area began at Amsterdam in 1997, in the same Treaty that 'founded' the EU on specific values. Here, EUrope (without the UK, Ireland and Denmark who negotiated opt-outs) committed itself to adopting measures on immigration, asylum, refugees and displaced people as part of creating EUrope as an Area of Freedom, Security and Justice (AFSJ), protecting the Schengen zone of border-free mobility. The AFSJ was firmed up by the Tampere European Council (European Council, 1999a), which laid the foundations for a common asylum and migration policy. Progress was made through three five-year programmes (Tampere, 1999; Hague, 2004; Stockholm, 2010), and the Lisbon Treaty, agreed in 2007, brought residual areas under supranational law-making. EUrope committed itself to 'a common policy on asylum, immigration and external border control, based on solidarity between Member States, which is fair towards third-country nationals' (Article 67(2) TFEU). And it was in the face of the mounting refugee crisis that a European Agenda on Migration (EAM) was finalised in 2015 (European Commission, 2015a).

While EUrope as a space of values was formalised alongside its migration policy from the late 1990s, enlargement has been an explicitly hospitable production of EUrope as a post-sovereign ethical space since 1993. 'Enlargement' refers to the extended process by which states apply for membership of the EU, become 'candidates' and negotiate their entry as member states. EUrope has been through seven enlargements, each of which has transformed and extended it. From its original six members, it welcomed the UK, Ireland and Denmark in 1973, Greece in 1981, Spain and Portugal in 1986, Austria, Finland and Sweden in 1995. The 2004 'big bang' enlargement saw the entry of ten new countries,[5] with Bulgaria and Romania joining in 2007 and Croatia in 2013. Negotiations for the latter three enlargements began while EUrope was institutionalising its founding values, making them particularly significant to this chapter. Even now 'the "waiting room" is far from empty' (Füle, 2013b): Turkey, Montenegro, Albania, Serbia and Macedonia are all 'candidate' states; Kosovo and Bosnia Herzegovina are considered 'potential candidates'.

Enlargement has been hailed as EUrope's most successful foreign and security policy by politicians and academics (Patten, 2005: 152; Phinnemore, 2006: 7; Füle, 2014a). It has helped protect the home, proving the 'best way to ensure the long term security of Europe' (Rehn, 2006f), immunising it from the threatening instability and insecurity of the Balkans and post-Soviet Central and Eastern Europe (European Council, 1993: 7.A.ii; Füle, 2010d, 2014a). It is also considered a moral obligation and responsibility (Patten, 2000c; Solana, 2000a, 2000b; Prodi, 2002c; Rehn, 2009a; Füle, 2010c), a hospitable expression of EUrope's ethos in relation to an often threatening outside world (Rehn, 2005a). Enlargement is spatially characterised by EUrope through 'opening doors' and 'welcoming' the other inside (European Council, 2000, 2002b, 2003, 2004, 2011, 2014; Prodi, 2002d; Solana, 2003;

Rehn, 2005a, 2006a, 2007, 2008d; Füle, 2010a, 2011b, 2014b). A great stress is also placed on the affective sense of belonging necessary to entry. Becoming a member of the EU is not only about economics, 'it is first and foremost a sense of belonging. Belonging to the European family, belonging to a community based on the rule of law' (Patten, 2000a). It is about countries 'destined to join' (European Council, 1997: 10, 1999b: I.12), that have 'returned to the European family' and are looking for 'the rest of Europe to welcome [them] home' (Patten, 2000a). It is a matter of states that *belong* in EUrope – part of the family, sharing its values – being officially welcomed inside.

This is recognised in Article 49 TEU which establishes the basis for enlargement: 'Any European State which respects the values referred to in Article 2 and is committed to promoting them may apply to become a member of the Union.' In other words, to be eligible for EUrope's hospitality, you must already be a 'European State'. To become EUropean, a state must not only already be European, but also EUropean, internalising the values that constitute belonging. EUrope can only welcome its self. Indeed, EUrope actively constitutes and immunises itself through its practice of hospitality, transforming its geography, territory and borders. Thus, Prodi argued that the 2004 enlargement would generate 'a new structure for our common European home' (2002b). As its space is constituted by values, spreading those values and welcoming in states that share them shifts its borders, whilst also making it more secure. Thus, 'successive enlargements of the EU have made it what it is today' (Rehn, 2006d). The next section will examine the process and power relations involved in EUrope's most successful form of protective hospitality – the road to a state's membership.

## THE ROAD TO EUROPE: IMMUNITY, CONDITIONALITY AND ENLARGEMENT

The conditions placed on EUrope's hospitality are easily the most thoroughgoing of any explored in this book, reflecting its focus on guarding the home from threat and external corruption. It is therefore not a 'natural' immunity that EUrope seeks, but an 'acquired' immunity that involves taking into the community or body a small amount of that which endangers it. An 'acquired immunity' thwarts a threat 'not by keeping it at a distance from one's own borders; rather, it is included inside them ... The body defeats a poison not by expelling it outside the organism, but by making it somehow part of the body' (Esposito, 2011: 7–8). EUrope's immunising hospitality works slightly differently, however, welcoming the outside as the final stage of a process that has purged it of all threat. As such, it is a peculiar hospitality. The spaces and assembled hosts examined in previous chapters have exercised power and control over the stranger at the threshold and once they are inside the home, seeking to manage the way strangers circulate and behave.

In contrast, EUrope's hospitality to states is exercised *before* they enter, because once they are welcomed they are *no longer strangers*. Power is thus concentrated on the 'road' to EUrope, a liminal space neither fully inside nor outside the home. Once completely inside, the idea (if not the reality)[6] is that entrants are treated the same as any other member state. But what sort of an immunising power relation explicitly seeks to 'transform' the stranger, stripping them of their threatening strangeness? To answer this question, we initially need to specify the immunising conditionality of EUropean hospitality.

## STRICTER CONDITIONS, LONGER ROAD, GREATER IMMUNITY

The conditions of EUrope's hospitality are expressed, first, in the so-called Copenhagen Criteria mentioned above. Eligible applicants, as well as being European states that respects EUropean values, must have a 'functioning market economy as well as the capacity to cope with the competitive pressure and market forces within the Union' (European Council, 1993: 7.iii). A second paragraph adds: 'The Union's capacity to absorb new members, while maintaining the momentum of European integration, is also an important consideration in the general interest of both the Union and the candidate countries' (ibid.: 7.iii). Conditions are here placed on the host as well as the neighbour. The body must be able to 'absorb' the 'poison'; otherwise the poison may absorb the body. Rehn (2006e) puts it more hospitably: 'every time we welcome a new member to our family, we want to ensure that the house is comfortable and functional for everybody'.

The application to join the EU is just the first step on what is characterised as the 'journey', or 'road', to EUrope. This 'road' is a liminal immunising space, characterised as 'long' (Rehn, 2006f; Füle, 2011b), 'hard' (Ashton, 2010a) and 'difficult' (Füle, 2013d), with 'staging posts' and benchmarks to be met along the way (Patten, 2000b), 'paved with concrete reform, not just good intentions' (Solana, 2003; Rehn, 2008a). It included detailed 'roadmaps' for the more problematic Romania and Bulgaria (European Commission, 2002; European Council, 2002a). Crucially, this 'road' has also become longer and harder; each enlargement has produced new conditions and their stricter application. Once the Commission has judged the fulfilment of these criteria through examining an aspirant's answers to an elaborate questionnaire, evolving practice dictates that the European Council must decide whether to determine them a 'candidate'. Here, additional criteria can be set before negotiations begin. For the Western Balkans, this meant signing and implementing Stabilisation and Association Agreements (SAAs, mirroring the 'Europe Agreements' with CEE countries) which contained both general requirements, such as the establishment of a free trade area with the EU, and more specific issues, such as the return of refugees and compliance with the International Criminal Tribunal for the Former Yugoslavia (ICTY) (Phinnemore, 2003; Pippan, 2004).

Should member states in the European Council give the green light, the first stage in the negotiations proper is a 'screening' process. Like a health screening, the candidate is subjected to a fine-detailed scrutiny by the Commission to determine where its deficiencies lie in terms of living up to the *acquis communitaire* (the body of EU law) and the obligations of membership. As the EU's areas of legal competence and regulations have increased with each treaty, so has the *acquis*, helping to elongate and complicate the road. The accession negotiations proper involve the splitting of the *acquis* into 35 different chapters, each dealing with a specific EU policy area (free movement of goods, capital and labour, energy, transport and regional policy, etc.) to ensure readiness for membership. Few of these have much to do with values; most are bureaucratic requirements for joining the single market. Each chapter contains benchmarks and progress must reach a certain level before a new chapter is opened. Whereas in the past compliance was required *at the point of entry* into the EU, after 2004 compliance is required *before* a chapter is closed, and a good track record of that compliance demonstrated before accession (Phinnemore, 2006: 18). Immunity must now be a sure thing.

Given this changing conditionality, the longer you stay on the road, the harder your journey becomes. Thus, Croatia became the EU's newest member in July 2013, having applied over ten years previously (February 2003) and begun negotiations in 2005. Turkey applied for membership of the EEC in April 1987, was recognised as a candidate in 1999, with the 'screening' process only starting in 2005. Since then, 14 of the 33 chapters requiring negotiation have been opened, 16 are frozen, with only one having reached closure. Nearly 30 years after its application, Turkey is not much nearer EUrope's door. It remains too poisonous to be fully absorbed.

Sovereign power, considered as the decision to welcome or exclude, rears up several times along the road to EUropean hospitality. For instance, the European Council must unanimously approve the Commission's recommendation to begin accession negotiations, meaning each member state must agree to open the road. Repeated Commission requests to open negotiations with Macedonia have been blocked by Greece, not because of its readiness but due to a dispute over its use of the name 'Macedonia'. Furthermore, once all the conditions of EU membership have been fulfilled and every chapter is closed, the final terms of accession are set out in an accession treaty which must receive the support of the European Council, Parliament, the candidate and every single member state. Yet, while this sovereign power appears key to EUrope's hospitality, it is only one step on the road, and not a particularly significant one. As the threshold is reached, the 'decision' becomes a formality; it has effectively already been taken, subsumed within the negotiations, the series of smaller decisions to open and close chapters of the *acquis*. No candidate has yet reached the end of the road and been denied entry, though since 2004 it has been stressed that negotiations are an 'open-ended process, the outcome cannot be guaranteed beforehand' (European Council, 2004: 23). Thus sovereign power emerges only at certain points within a general field of governmentalities.

## THE PASTORAL ETHOPOLITICS OF THE ROAD

Foucault notes that one of the many meanings of 'to govern (*gouverner*)' is the spatial sense of 'to direct, move forward, or even to move forward oneself on a track, a road. "To govern" is to follow a path, or put on a path' (2007: 121). This understanding is closely tied to one of its earliest incarnations, discussed in Chapter 2: pastoral power, exercised over 'a flock ... a multiplicity in movement' (ibid.: 125). Similarly, while EUrope is working to immunise itself by placing states on a 'road' to EUrope, it defines its relation to these states as that of the shepherd guiding a flock. Since the 1990s, enlargement has been setting a multiplicity of countries on this road, first the CEE (plus Malta and Cyprus) flock who joined in 2004 and 2007, and subsequently the Western Balkan flock of seven countries (plus Turkey). Crucially, the relation between the shepherd and the flock is hierarchical: the two are not equals (Foucault, 2007: 124). As Rehn (2005d) clarifies:

> The negotiation process for Turkey means nothing more or less than Turkey adopting the rules of governance which are applied in today's Europe ... In this sense, the word negotiation here is perhaps misleading; the discussions will in fact focus on 'how' Turkey will adopt European standards and not on 'whether' Turkey will adopt them. One of the fundamental principles of EU membership is that candidate countries must adopt all of the EU's laws and policies.

The language of 'negotiation' is used to efface the hierarchical power relation between the shepherd (EUropean institutions) and the flocks seeking to accede.

Yet, at the same time, this process is not one of domination and coercion. Enlargement is about freedom and choice – it is entirely 'voluntary' (Solana, 2005). With the possible exception of Kosovo (Musliu, 2014), EUrope does not force any state to apply for membership, nor does it force its reforms upon them. Resistance is both possible and, for some, simple: Switzerland, Norway and Iceland have all applied to join the EU, but subsequently removed themselves from the road by either freezing their application (Switzerland), failing to ratify the Treaty of Accession (Norway – twice), or deactivating their application (Iceland).[7] EUrope seeks to lure and seduce Norway (Rehn, 2009b) rather than compel it. But resistance is asymmetrical: the economic benefits EUrope offers (access to the internal market, structural and investment funds, etc.) are easier for wealthy Northern Europeans to live without (or negotiate non-member access to) than relatively impoverished Romania and Albania.

A second aspect of pastoral power is that it is defined by 'benificence' – its 'essential objective ... is the salvation (*salut*) of the flock'. The shepherd's role is to feed and secure the flock, thus it is a 'power of care ... it goes in search of those that have strayed off course, and it treats those that are injured' (Foucault, 2007: 126–127). The caring nature of EUropean power is partly demonstrated through its very hospitality – the home, with its peace, stability, prosperity and liberal values, is being opened to those who have experienced their opposite: anarchy, war, instability, poverty and authoritarianism. EUrope must immunise itself against the threat that such openness poses, but the neutralisation of that threat is also a caring, beneficent relation. Recent enlargements have been about 'a region escaping from 45 years of totalitarian government and neglect' and EUrope exercising a pastoral responsibility to 'help them complete that journey' (Solana, 2000c). As a good shepherd, this includes 'help[ing] the straggler along'.

The third element of pastoral power is that it is both massifying (caring for the flock as a whole) and *individualising*; the shepherd must 'keep his eye on all and on each' (Foucault, 2007: 128). The accession process set out by the European Council in 1997 (i.14) called for a 'single framework' for the flock, but promised that each state would be guided 'individually'. The shepherd is one who watches over the flock to 'avoid the misfortune that may threaten the least of its members' (Foucault, 2007: 127). Thus, the Commission 'monitors closely developments in the countries, and reports on both progress and shortcomings', whilst assisting each individually, 'both financially and with policy advice' (Rehn, 2008b). The result is that individual candidates are cajoled to 'catch up' with the rest of the flock and front-runners are praised as an 'example' for others to follow (European Council, 2000: I.D.15): Serbia was considered a 'straggler' in 2000 (Solana, 2000c); 14 years later it could be an 'example to others' (Ashton, 2014). The consistent straggler demanding the most watching has been Bosnia. Even when designated a Potential Candidate in 2000, External Relations Commissioner Chris Patten saw Bosnia's progress as too slow (Patten, 2000b). Ever since, it has been identified as dragging its feet and 'risks being left behind [by] the other countries in the region' (Rehn, 2009f; see also 2006c; Patten, 2001, 2004; Ashton, 2010a; Füle, 2012a, 2013c).

Looking more closely at EUrope's relation to the Bosnian 'straggler' reveals that its immunising pastoral power is supplemented by a more subtle form of what Nikolas Rose called 'ethopolitics'. Rose identifies 'ethopower' as operating through the way 'community' and its values, norms and way of life (its ethos) are being reformulated and instrumentalised in advanced liberal Western societies as a way of 'governing at a distance' (Rose, 1999, 2000a, 2001). A form of pastoralism, ethopower works more through the relation between the ethics, values and affects of the guider and those of the guided (Rose, 2001: 9), making it peculiarly applicable to EUropean hospitality. Rose specifies that while 'discipline individualises and normalises, and biopower collectivises and socialises, ethopolitics concerns itself with *the self-techniques by which human beings should judge themselves and act upon themselves to make themselves better than they are*' (Rose, 2001: 18, emphasis added). The immunising logic of EUrope's hospitality, and the purpose of its 'road', is explicitly about invoking such a transformation. The EUropean home provides the ethical benchmark and immunised destination; the road's screening and negotiations are about supplying the tactics and techniques by which candidates can judge and act upon *themselves* in order to make themselves better: *more* liberal, *more* democratic, *more* respectful of human rights, *more* EUropean. It is to encourage such self-betterment that EUrope is putting greater emphasis on the rule of law and other 'softer' aspects of the *acquis* in its negotiations (Grabbe, 2014). EUropean institutions will judge their success, but what makes the primary dynamic ethopolitical is that the actual work of the road is performed *on* the acceding self and *by* the acceding self, to make itself *better* (less threatening, less poisonous).

One of the key tactics of ethopolitics, that which separates it from a strict pastoralism, is what Rose calls the 'double-movement' of autonomisation and responsibilisation – those once directly controlled and governed are 'set free to find their own destiny. Yet, at the same time, they are made responsible for that destiny' (Rose, 2000a: 1400). Candidates are judged and judge themselves on the extent to which they live up to and achieve that destiny. Solana reflected at the end of his long term as High Representative that the immunisation dilemma EUrope faced in the Western Balkans was precisely that of autonomy or tutelage, offering enlargement or a 'protectorate of sorts' (2009b). EUrope chose to offer autonomy and enlargement, 'conditioned on reform'. However, the ethopolitics of the road is most intensely focused on the extreme case of Bosnia, which has proven unwilling to make the *right* choice and fully accept its responsibility to become *better*. Addressing a primarily Bosnian audience, Commissioner Patten (2000b) stressed the fact that autonomy required Bosnians to take responsibility:

> We have to redouble our efforts and focus our attention on the really urgent priorities. I say *we* in its most inclusive sense. But it is a we whose main burden actually falls on you, you the leaders and people of Bosnia and Herzegovina. It is for *you*, not principally for the EU or the international community, to put your country on the road to Europe; it is for you to set the pace, for you to determine how rapidly you arrive at your destination. We can point the way, as we have done through our EU road map of measures we want to see you fulfil before embarking on the stabilisation and association process; we can help build that road, as we are doing through our very substantial assistance … we can encourage and assist you every step of the way. We can and we will – ensure BiH [Bosnia and Herzegovina] never has to walk the road to Europe alone. But we cannot carry you the whole way along it.

The shepherd is watching the weak sheep, but denies ultimate responsibility for its fate; responsibility is shifted to the sheep – Bosnia and its people. As Rehn later stressed, 'We cannot travel the road to the EU for Bosnia and Herzegovina' (2009c). It is Bosnians themselves who are accountable if they remain 'outside, in the cold' (Füle, 2013c). It is up to the candidate countries to demonstrate their belonging on the road to EUrope, the fact that they are not only capable of being governed via a EUropean ethos, but that they are already *governing themselves* according to these principles. They are responsible for the immunisation of EUrope. This is why a good track record of *acquis* compliance must now be demonstrated *before* accession.

It is in terms of an immunising pastoral ethopolitics that we can perhaps best understand the way Enlargement Commissioners have consistently referred to EUrope's 'transformative power' (Rehn, 2008c, 2009g; Füle, 2011c, 2012b, 2014a).[8] The purpose of 'the road' in EUrope's hospitality is explicitly that of transforming the subject from non-belonging to belonging, from destabilising to stabilising, from poison to cure. What is understood by 'transformative power' is a combination of the EU's 'gravitational pull' alongside stricter conditionality (Rehn, 2004). These tactics changed the CEE countries into 'modern, well-functioning democracies' and are now transferred to the Western Balkans (Rehn, 2006b). Thus, when they accede, the flock will be 'transformed into the kind of neighbours we would like to have – stable, secure, well-governed and prosperous ... fully part of mainstream Europe' (Rehn, 2005d), family rather than neighbours. EUrope will be immune to their threatening difference because they will no longer be different. However, while the notion of 'transformative power' is revealing, it is also tautological (all power is productive, and thereby transformative) and fails to account for the ethopolitics of the road – the fact that the road to EUrope, if perfectly constructed, allows candidates to govern, neutralise and transform *themselves*.

## EUROPEAN PROTECTION: MIGRATION, ASYLUM AND OUTSOURCING

As we have seen, the immunising ethopolitical conditions placed on EUrope's welcome are unusually restrictive compared to the spaces considered in previous chapters. But it also has 'higher' aims than the humanitarian protection of the refugee camp (see Chapter 2), the indifferent flourishing of the global city (Chapter 3), or the now commercialised near-unconditionality of the postcolonial state (Chapter 4). It appears closer to the idealised family home of *Welcome to Sarajevo* (Chapter 1), especially in its pastoralism. However, this ethopolitical hospitality also demands a non-threatening subject, transforming itself to become worthy of welcome. It must purge its difference; already *belonging* inside it must be a modern, liberal, democratic European state, with a track record of respecting and upholding EUrope's values.

I now turn to a more controversial EUropean practice of hospitality – immigration and asylum policy. While the 'threat' posed by states emerging from authoritarianism, civil war and ethnic conflict has also been interpreted as an opportunity for EUrope (in terms of economics, security and ethics), irregular migration has been more consistently portrayed as a threat to the EUropean home (see Huysmans, 2006). There are frequent calls for greater hospitality towards the *right kind* of immigration as a necessary supplement to EUrope's ageing population and labour shortages (Frattini, 2005d; European Council, 2006; European Commission, 2011: 12–13, 2015a: 14–15; Avramopoulos, 2014c; Juncker, 2015b). In contrast, irregular migration is something that must be 'fought' as a threat to EUrope's labour markets, social cohesion, welfare systems and governance practices (Vitorino, 2001; Frattini, 2005a, 2005b, 2005c, 2005d, 2006b; European Council, 2006, 2009, 2015c; Malström, 2010a, 2010b, 2013b; European Commission, 2011: 2–5; Avramopoulos, 2014a).

Yet, the solution to this threat is not a closing of doors. Rather, it is better management (European Council, 2015a) – thus the 2015 EAM proposes four 'pillars' to 'manage migration better' (European Commission, 2015a: 6). Immunisation is not to be achieved through making EUrope into an impregnable fortress, but by taking a small amount of the threat inside – '[i]t reproduces in a controlled form exactly what it is meant to protect us from' (Esposito, 2011: 8). In doing so, EUrope's hospitality seeks to protect *both* the EUropean home *and* the migrants and refugees themselves – the EAM's better management is about 'saving lives' as well as 'securing external borders' (European Commission, 2015a: 10–11); the Commission's 2011 Global Agenda on Migration and Mobility (GAMM) is 'migrant centred' and makes protecting the human rights of migrants a 'cross-cutting dimension' (European Commission, 2011: 6). This stress on protection means EUrope's migration policy is difficult to criticise from a conventional humanitarian perspective as it has co-opted the discourse of humanitarianism (Vaughan-Williams, 2015). The protection of migrants, particularly refugees, is central to EUrope's immunising hospitality. It operates via a pastoral biopolitics that, like the 'road', creates liminal spaces of protection which are *becoming* EUropean, for ever inside and outside its space but never entirely either.

## MIGRATION, MOBILITY AND ASYLUM: EXCLUSION AND CONDITIONALITY

Immigration and asylum is still an emerging area of (in)competence for EUrope. The aim of cooperating on migration and asylum from 1999 was both a reaction to pressures generated by the free movement of people under Schengen *and* part of a broader attempt to create EUrope as a particular kind of space: 'an area of freedom, security and justice' (European Council, 1999a). Crucially, while 'freedom' and 'justice' are 'founding' values of EUrope's ethos, security stresses the immunising logic of this hospitality. Its aim was not unqualified welcome, rather it would be primarily *protective*: a careful, watchful hospitality. Former Vice-President of the Commission Franco Frattini thus claimed that EUrope has 'two very different faces' on migration, depending on its legality (Frattini, 2005b). The EUropean home aimed at 'closing the back door firmly whilst opening the front door of legal migration' (Vitorino, 2001).[9] The liminal subjectivity of the 'asylum seeker', neither legal nor illegal, would be caught by the emerging Common European Asylum System (CEAS).

In welcoming legal migrants, EUrope's concentration is on its economic progress and ageing working population. Here, it attempts hospitality comparable to the global city (Chapter 3), reversing the trend of EUrope receiving 'low-skilled or unskilled labour' while the US, Canada and Australia 'are able to attract talented migrants' (Frattini, 2006a). The aim is therefore 'to attract the smartest and the brightest', to 'improve the attractiveness of the EU as a destination for highly qualified migrants' (Malström, 2012) by offering 'new European employment possibilities for talented people from around the globe' (European Commission, 2011: 12). Legislation in this area includes the 'EU Blue Card Directive', which offers to welcome the highly qualified on a temporary basis (Council Directive 2009/50). Though EUrope mimics the hospitality of the global city, it has also shown an awareness of its global responsibilities that are completely effaced by London's pursuit of the talented. Wary of the 'brain drain' effect on developing countries, EUrope tries to support 'brain circulation', or 'circular migration' (Frattini, 2005c, 2006b, 2007a), through Mobility Partnerships (Commission, 2011: 12) and 'ethical recruitment' outside certain strategic sectors of developing countries (Council Directive 2009/50, Article 5(3)).

Cooperation promoting legal migration has, however, proven 'nascent and weak', with achievements 'far less grand' than Commission proposals (Geddes, 2014: 447). Greater successes are evident in areas that restrict migration rather than enable it, feeding criticisms of 'fortress Europe' (Lahav, 2014: 458; Hansen, 2009). Closing the back door to illegal migration has seen both more agreement and implementation. Most prominent in this was the inauguration of Frontex in 2004 to establish, coordinate and oversee 'integrated management of the external borders of the Member States of the European Union' (Council of the EU, 2004: Article 1(1)). Boosts in funding and changes to its mandate have seen Frontex take an increasingly militarised role, using drones, aircraft, offshore sensors and satellite technology to track and trace illegal migration into EUrope through the Eurosur surveillance system. Meanwhile, Directives and strategies have been agreed on the return of irregular migrants, trafficking and sanctions against employers who use irregular workers (European Commission, 2015a: 15–17). 'Europe has declared war against smugglers', according to the Migration Commissioner (Avramopoulos, 2015), with Common Security and Defence Policy operations being proposed to 'systematically identify, capture and destroy vessels used by smugglers' (European Commission, 2015a: 3).

The violence of this hostility has been tempered by the humanitarian mandate given to Frontex, which sees it saving migrants and protecting their rights as well as securing borders (European Commission, 2011, 2015a; Walters, 2011; Vaughan-Williams, 2015). EUrope's hospitality to migrants can therefore only be reductively portrayed as, on the one hand, an economically driven, heavily conditional and largely ineffective welcome; and on the other, a security driven, disciplinary and militarised hostility. While economic prosperity and security are values of a sort, and appear central to what EUrope is becoming, neither are claimed as foundational or essential. They are not central to the ethos of the EUropean home. Where EUrope's ethos of hospitality seem to emerge most clearly is in its 'humanitarian government of migration' (Walters, 2011: 146), encompassing the humanitarianisation of the EUropean border (Vaughan-Williams, 2015) and the movement towards a common asylum system (CEAS). The latter particularly targets the liminal category of asylum seekers whose legality is not yet determined; she may arrive by regular or irregular means, but has a human right to do so.

Since the Tampere European Council of 1999 (A.II.13), EUrope has been slowly building and implementing the CEAS. This has been given significant prominence, with Commissioner Cecilia Malström (2013a) making it the 'top priority' of her Home Affairs mandate from 2010–14. The aim of the CEAS is that of making EUrope an 'Area of Protection' (Barrot, 2008; Malström, 2010b; European Commission, 2014), based on the 'common values underpinning the Union' (Frattini, 2005a; European Council, 2006). Asylum is thus the area where EUrope most clearly expresses its ethos in relation to the non-state other coming from outside the home. And it does so through the humanitarian government of hospitality. It is here that EUrope is produced not only as an area of 'freedom, security and justice', immunising the EUropean home and its citizens, but also as an 'area of protection' for those arriving at its door.

## PROTECTING THE STRANGER: PASTORAL BIOPOLITICS OF IMMUNISATION

For such a noble, protective cause, the CEAS itself is largely uninspiring. Its first phase (1999–2005) concentrated on harmonising member states' legal frameworks on asylum around minimum standards (European Commission, 2008b, Annex II). The Hague Programme set up the second phase whereby a common asylum procedure would ensure the speed, efficiency and fairness of decisions; a uniform status for those granted protection would be guaranteed; greater administrative cooperation between member states on training and burden sharing; and concentration on 'the external dimension of asylum' (European Commission, 2008b: 2–3).[10] It is this 'external dimension' where most innovation has occurred. Yet, even now the CEAS remains incomplete as it is not adequately and uniformly implemented (Malström, 2013a). Thus, from 2014, the focus shifted to monitoring and ensuring implementation (Malström, 2014a), though huge difficulties surround proving violations by member states: 'claims are not always dealt with in the light of a court room' (Malström, 2014b).

This attempt to create EUrope as a space of protection is made through the combination of pastoralism and biopower which Fassin (2012) calls 'humanitarian government'. Like the refugee camp (Chapter 2) and the 'road' of enlargement, EUrope exercises a power of care over a multiplicity in movement, guiding them to safety and protection. While there is rarely a mention of EUropean values or ethos in the CEAS, its aim is to be worthy of 'our European humanitarian traditions' (Malström, 2011b), encompassing its entire ethos. Yet, this power of care is entirely massifying, targeting the population of asylum seekers, saving their lives (and letting them die) (Foucault, 2004: 240–243). Its interventions are thus biopolitical rather than ethopolitical: there is no attempt to concentrate on the individual or manage her ethos; interventions are made in the lived existence of the refugee and host populations.

Thus, as the refugee crisis worsened in 2015, the Commission, building on CEAS proposals, suggested two mechanisms for offering immediate protection (European Commission, 2015a: 4–5, 2008b: 10–11). These were a 'relocation scheme', where member states overburdened by arrivals would have refugees and asylum seekers removed to other territories using a 'redistribution key based on criteria such as GDP, size of population, unemployment rate' and existing refugee population; and a 'resettlement scheme' offering 20,000 places based on similarly weighted 'objective, quantifiable and verifiable criteria that reflect the capacity of the Member States to absorb and integrate refugees' (European Commission, 2015a: 19). EUrope as a space of protection is therefore governed by a rationality more akin to the humanitarian hospitality of the refugee camp than the ethopolitical hospitality of enlargement. But both enlargement and asylum policies are concerned with the capacity of EUropean space and its communal body (or population) to 'absorb' their poison. The key difference between the humanitarian hospitality of the refugee camp and the immunising hospitality of EUrope lies in the 'external dimension' of the CEAS, which has become its increasing focus (Eurpean Commission, 2008b: 9). Where the idealised refugee camp sets up a system of way-stations and transit centres, forming and guiding a population to protection, the external dimension of EUrope's protection does not facilitate movement. Quite the opposite. Thus, the problem for those seeking the protection of EUropean space is precisely how to reach it. As Malström (2014a) observed, 'asylum seekers have to rely too often on traffickers in order to reach Europe. There are basically no legal ways to get to Europe.'

The central feature of the 'external dimension' of EUropean protection lies in its production and transformation of space:

> The EU must share the responsibility for managing refugees with third countries and countries of first asylum, which receive a far greater percentage of the world's refugees than Europe. In this regard, more financial

> support will be available to *enhance protection capacity in third countries* ... Furthermore, the Commission will continue to integrate capacity building for asylum in development cooperation with third countries, placing the emphasis on a *long term, comprehensive* approach. Asylum should not be treated as crisis management but as [an] integral part of the development agenda in the area of governance, migration and human rights protection. (European Commission, 2008b: 9, emphasis added; see also European Council, 2002a, 2006, 2009, 2015b)

EUrope's protection is offered by supporting *other* spaces and territories. Like international aid, and indeed often as a part of it, EUropean protection can be 'delivered' outside the home (Frattini, 2005c). The external dimension of asylum is thus central to both the GAMM (European Commission, 2011: 17–18) and the EAM (European Commission, 2015a: 7–10). As 'external', these policies appear disconnected from EUropean space, 'interven[ing] upstream in regions of origin and of transit' by increasing financial and other forms of 'support' (ibid.: 5), cooperating to 'strengthen' these countries' asylum systems and legislation (European Commission, 2011: 17; European Council, 2006, 2009). Migration control is now included in *every* trade, development or security agreement with third countries: as early as 2002 the European Council urged that 'any future cooperation, association or equivalent agreement ... with any country should include a clause on joint management of migration flows and on compulsory readmission in the event of illegal immigration' (European Council, 2002a: 33).

The most concrete example of this apparent 'outsourcing' of protection (Gammeltoft-Hansen, 2011) is the greater use and development of Regional Protection Programmes (RPPs), which 'reinforce the external dimension of asylum' (European Commission, 2008b: 10, 2011: 17, 2015a: 5; Barrot, 2008; Malström, 2011d). RPPs are a matter of 'enhancing the protection capacity' and promoting durable solutions in countries of origin and transit through protection, reception, registration and status determination training as well as financing (European Commission, 2005: 2–4). The first two RPPs targeted Eastern Europe as a transit region (Belarus, Moldova and Ukraine) and the African Great Lakes Region (particularly Tanzania) as a region of origin. A second wave was set up after 2010, covering the Horn of Africa (Kenya, Yemen and Djibouti) and North East Africa (Egypt, Libya and Tunisia). In 2013 a Regional Development and Protection Programme for Syrian refugees was announced, covering Lebanon, Jordan and Iraq (European Commission, 2013). RPPs therefore now surround the EUropean home, raising the standard of protection in states which produce and transit refugees.

However, along with humanitarian care and protection, RPPs work to *contain* refugees by criminalising their further movement. They fulfil a dual purpose: to 'ensure that those who need protection are able to access it as quickly as possible and as closely as possible to their needs', whilst also 'prevent[ing] illegal secondary movements' (Frattini, 2005c; see European Commission, 2005: 6 fn. 4).

Though RPPs retain a 'resettlement commitment' from EUropean states, this is 'on a voluntary basis' and therefore commits to nothing (European Commission, 2005: 4). While RPPs may appear like transit centres for refugees fleeing to camps, their function is not to facilitate onward movement but to *arrest* it, to *halt* transition and *block* the road to EUrope. This approach was supplemented in response to the current refugee crisis in November 2015 as agreement was reached with African leaders at Valletta on an 'Emergency Trust Fund'. This will involve over €1.8 billion of development aid for countries in North Africa, the Horn of Africa, the Sahel and Lake Chad regions in return for 'addressing the root causes of irregular migration and promoting economic and equal opportunities, security and development' (Juncker, 2015c).

The Valletta agreement was, for some, a foregone conclusion, merely entrenching an existing 'politics of inhospitality, denial of basic rights and cynical bargaining' (Blanchard et al., 2015). Yet, hospitality and hostility (or inhospitality) cannot be so easily opposed (Introduction, Chapter 4). Any condition placed on an open welcome constitutes a hostile negation of hospitality, yet an unconditional hospitality would destroy the home which is the condition of hospitality. The CEAS and its 'external dimension' is thus a negotiation of hostipitality. Rather than simply hostile, the 'external dimension' of asylum is better read through the immunising pastoral biopolitics by which EUrope practises its hospitality. Instead of welcoming refugees into the EUropean home, it seeks a minimal transformation in areas that produce and transit refugees, making them more EUropean. It is thus important to recognise 'the ways in which the exercise of humanitarian power is connected with the actualization of new spaces' (Walters, 2011: 139). The regions of protection that surround the home are not just *funded* by EUrope, the aim is to transform them into *humanitarian* spaces, making them *more EUropean*: to raise their standard of protection to international and EUropean standards, including respecting the rights of migrants through training and legislative reform.

This transformation is interpreted as an expression of solidarity, a value central to the EUropean ethos; solidarity with the 'developing world' (European Commission, 2008b: 9; Malström, 2014b) and 'solidarity with refugees and displaced persons' (European Commission, 2011: 17). Though EUrope consistently fails to define what it understands by 'solidarity', it can be etymologically traced to the Latin adjective *in solidium*, meaning 'for the whole' (Hoelzl, 2004: 51). Expressing solidarity is therefore articulating something as 'whole', as complete, rather than separate and distinct. As we saw above, the space and limits of EUrope are fuzzy and defined by values; if the values of justice, human rights, solidarity and the protection they result in are exported through RPPs and aid, these spaces are no longer *simply* external. They are external to the territory of the EU-28, but not to EUrope and its ethical space, which only ends where its values are no longer shared. These 'external' spaces thus become part of a whole, part of EUrope as an 'area of protection'; they are where most of EUrope's protection is delivered, on the basis of shared values through a raising of standards. Protection is thus part of the broader means by which EUrope, like a mediaeval empire, extends its space, regulations, rules and values through trade, development and cooperation (Zielonka, 2006). Such spaces are *becoming*-EUropean, even if they will never be welcomed as member states. They are a part of a minimalist, immunising and protective hospitality, a diluted form of enlargement rather than its opposite.

The two policies by which EUrope practises its immunising hospitality – enlargement and immigration/asylum – are united by a similar ethical and governmental logic. Just as states on the ethopolitical road to EUrope must accept the requirements and regulations of the CEAS (as part of the *acquis*) and practise biopolitical hospitality, so non-EU spaces are becoming-EUropean when they are encompassed by an RPP or other strategic partnership. We can certainly contest the morality and efficacy of the CEAS's 'external dimension' and its 'outsourcing' of protection, but it is also important to recognise that this is an outsourcing of EUropean values and space. While this operates to *prevent* the refugee making it to the borders of the EU-28, it also offers restrained EUropean protection according to EUropean values. Like the refugee camp, EUrope as an 'area of protection' exercises a pastoral biopolitics of care *and* control, its protective care being dependent on refugees' willingness to be controlled and excluded from the EU. But it is not only refugees that are being protected through this hospitality. As an immunising hospitality, the CEAS protects the home, its community and ethos by welcoming with minimalism.

## EUROPE'S QUASI-SUICIDE: AN AUTOIMMUNE ETHOS AND HOSPITALITY

Resistance to EUrope's immunising hospitality is not difficult to find. Despite Solana's (2005) claim that 'everyone wants to join this club and virtually no one wants to get out', Norway, Switzerland and Iceland have demonstrated otherwise. Some acceding states have joined while resisting elements of EUrope's conditionality, as David Phinnemore (2010) illustrates with the case of Romania. Meanwhile, refugees and migrants keep arriving by 'irregular' means, countering attempts to 'ensure that mobility and migration can be organised in an orderly fashion' (European Commission, 2011: 15). Looking at the 'routes' by which Frontex measures illegal arrivals, over 170,000 arrived through the 'Central Mediterranean' in 2014 (up 277 per cent on 2013) and over 50,000 through the 'Eastern Mediterranean' (up 104 per cent on 2013).[11] By November 2015, a further 128,000 people had arrived through the Central Mediterranean, while the Eastern Mediterranean saw nearly 360,000 arrivals.[12] A new record was reached, with nearly 220,000 people reaching Europe by sea in October alone (BBC, 2015c). None of these figures account for the main entry route, via legal means with visas that subsequently expire. Refugees and migrants are refusing to have their arrival governed through an immunising pastoralism, demanding a more fulsome protection and hospitality by 'visitation' without 'invitation' (Derrida, 2003: 128).

However, to concentrate on such counter-conducts makes it appear that the only threat to EUrope's immunising hospitality comes from outside: from resistant states and refugees. This could perhaps be remedied by even better management of EUrope's 'external dimension' and more seductive enticing of desirable states. My argument is rather that the most threatening resistance to EUrope is *internal*, that it is always already *inside* EUrope, contained in the ambivalence and contradictions of the home and the attempts to immunise it via practices of hospitality. In this sense, EUrope's hospitality is *auto*immune – it is 'quasi-suicidal' (Derrida, 2003: 94). The internal contradictions of its own ethos mean it menaces *itself*, its own values and protection, practised through hospitality. Autoimmunity is more threatening than a danger coming from outside; as an attack from the self on the self, it questions the very possibility of a 'self' (Derrida, 2005b: 45), endangering the fact or prospect of 'EUrope' *as such*.

Derrida's most overtly political reading of autoimmunity appears in his interpretation of the indeterminacy at the heart of democracy. This ambivalence emerges from democracy's privileging of two principles: freedom (which is necessarily incalculable and unconditional) and equality (which requires calculation and measurement) (Derrida, 2005b: 48). While democracy is unthinkable without people having the freedom to govern themselves, 'this freedom is immediately restricted within itself, since there is always more than one member of the people, which forces each one to act in relation to others that limit his or her freedom' (Hagglund, 2008: 172). Thus, democracy is also unthinkable without a calculative equality which imposes limits on an unconditional freedom: 'freedom is compromised by equality, and equality is compromised by freedom, but without such compromise there can be no democracy' (ibid.: 173). The condition of democracy is a constitutive autoimmunity: it is forever torn and divided against itself.

Derrida illustrates this with the example of Algeria in 1992. Faced with an expected electoral victory for an Islamic party pledged to end democracy, Algeria's democratic government cancelled the elections, to 'suspend, at least provisionally, democracy *for its own good* ... so as to immunize it against a much worse and very likely assault' (Derrida, 2005b: 33). Democracy necessitates the choice of allowing the freedom to murder democracy, or protecting and preserving the calculative equality of democracy by committing suicide; attacking its own protection in order to preserve itself. While an extreme example, Derrida's point is that this is *always* a necessary possibility contained within democracy. Thus the 'better management' of democracy cannot resolve its internal contradictions threatening it from inside. Complete freedom produces inequalities which restrict freedoms; calculating limits on freedom to preserve equality limits that which, democratically, cannot be limited. My suggestion is that a similar autoimmunity is at work in the EUropean ethos; its immunising hospitality always threatens to turn it into its opposite.

## THE DEMOCRACY OF MINIMALIST STATES

The autoimmunity of EUrope's enlargement hospitality can perhaps be best illustrated through the attempts to welcome so-called 'minimalist states' (Bieber, 2011) like Kosovo, Serbia (and Montenegro) and Bosnia Herzegovina. The Bosnian case is particularly revealing. EUrope has been heavily involved in Bosnia since the Dayton Peace Accords which imposed a weak central administration and split the effective government of Bosnia into a Croat–Bosniak Federation and the Serbian Republika Srpska (RS). There have been recurring calls for secession, especially from Bosnian Serbs who seek a reunification with Serbia (Biermann, 2014). The international community's presence in the form of the Office of the High Representative (OHR), which was to oversee the implementation of the peace, has barely held the 'state' together. Bosnia's extreme decentralisation and outside influence has meant that it continues to be a 'minimalist state' which 'barely fulfil[s] the functions generally associated with states' (Bieber, 2011: 1784; see Chapter 4). The Stabilisation and Association process, which began for Bosnia in 2000 as an early step on the road to EUrope, therefore insisted on a strengthening of the central government and an end to the OHR (Noutcheva, 2009: 1070–1071). This was not the usual 'institution building' of EUrope's pastoral ethopolitics; it was outright 'member-state building' (Rehn, 2005b, 2005c). The immunising nature of EUrope's hospitality *required* a stronger central state with which EUrope could negotiate (after all, EUrope can only welcome modern, liberal democratic states through enlargement), but which would also stabilise the region, curb ethnic tension and reduce insecurity in EUrope's neighbourhood.

Bosnia's 'partial compliance' with EUropean conditions on centralisation, for example on taxation and police reform, delayed the signing of the SAA until EUrope was satisfied in 2008 (Noutcheva, 2009: 1077). However, when demands for further centralisation and an end to the OHR, which would weaken RS and transfer powers to Bosniaks, was rejected by the democratically elected representatives of Bosnia (Noutcheva, 2012: 163–4), Commissioners expressed a frustration with democracy:

> Let me put it as plainly as I can: there is no way a quasi-protectorate can join the EU. Nor will an EU membership application be considered so long as the OHR is around. Let me even repeat this, to avoid any misunderstandings: a country with a High Representative cannot become a candidate country with the EU. It is a question of political maturity and leadership, not just a question of who sits at the table when we negotiate. (Rehn, 2009c)

However, minimalist states are not the 'result of ideological support for a state with minimal functions, but the consequence of a lack of consensus on endowing the state with greater competences' (Bieber, 2011: 1787). While RS was resolutely opposed to constitutional reform, Bosniaks were generally in favour and Croats held a shifting position between the two (Noutcheva, 2009: 1078). Bosnian democracy could not support the changes EUrope required. As such, EUrope's conditions of stability and security, which are at the heart of its hospitality as an immunising process, could only work by *attacking* democracy, a 'founding' value of the EUropean ethos.

This attack led to EUrope and the OHR pressing harder for constitutional reform and state centralisation in the 2000s, against the wishes of Bosnian Serbs. The effect was one of *fuelling* rather than dampening nationalist and secessionist parties, strengthening their position within their communities (Noutcheva, 2009, 2012; Bieber, 2011; Juncos, 2012; Biermann, 2014). EUrope's desire for long-term democratic stability thereby produced greater instability, bolstering the Serb nationalists it sought to oppose and weakening the position of Bosniaks and Croats seeking constitutional reform and EU membership. EUrope's hospitality was thus working against itself and its ethos, something Serb nationalists noted and utilised, denouncing the attempted impositions as undemocratic, 'unfair and against the European idea' (Noutcheva, 2009: 1079). Having fanned the flames of nationalism and separation, EUrope's response was to make it clear that secession was something it would 'never accept' (Ashton, 2010a), warning that RS 'can have as many referendums as it likes, but in the end, this is about one country coming together' (Ashton in Biermann, 2014: 501).[13] Democracy had to be protected by ignoring the democratic freedoms of Bosnian Serbs to decide how they are governed. Because of its autoimmune ethos, EUrope's pastoral hospitality is thus always in danger of reversing into its very opposite, what Grégoire Chamayou (2012: 11–18) calls the 'cynegetic power' of tyranny.

## SOLIDARITY WITH WHOM?

Unlike democracy, freedom and human rights, the undefined value of 'solidarity' has long been at the core of the EUropean ethos, included in the preambles to the treaties of Paris (1951) and Rome (1957). Yet, in the latter case, this was a 'solidarity which binds Europe'. Faced with rising numbers of refugees seeking EUrope's hospitality and protection, we saw above that this has been extended to include solidarity with refugees and 'third countries' (European Commission, 2008b: 9, 2011: 17; Malström, 2014b). But extending the space of EUrope in this minimalist manner has highlighted the autoimmunity of a EUropean ethos which consistently portrays its values, particularly human rights and freedoms, as both universal *and* specifically EUropean (Frattini, 2007b; Füle, 2011a; Avramopoulos, 2015). This has produced a battle, internal to the host, over the precise nature and terms of solidarity, human rights and freedoms.

Amongst the fiercest of these battles is between the Commission and member states. The Commission has consistently pleaded with the latter to show more openness and solidarity with migrants and refugees, as well as with each other. As Home Affairs Commissioner, Malström was especially clear on the need to share 'responsibility' for offering protection (2010a, 2011c, 2012, 2013b, 2014a). Her criticism peaked one year after the Lampedusa shipwreck, where over 360 refugees were drowned in one incident:

> Let me be very clear – when it comes to accepting refugees, solidarity between EU member states is still largely non-existent. This is quite possibly our biggest challenge for the future. While some EU members are taking responsibility, providing refuge for thousands of refugees, several EU countries are accepting almost no-one. In some countries, the number of yearly refugees barely exceeds a few handfuls. Last year, six whole countries of the EU accepted less than 250 refugees between them. All this, while the world around us is in flames. These EU countries could quite easily face up to reality by accepting resettled refugees through the UN system, but despite our persistent demands they are largely refusing. This is nothing short of a disgrace. If all the promises after the Lampedusa tragedy are to mean anything, solidarity between EU countries must become reality. For this to happen, we must in the coming years develop a responsibility-sharing mechanism between all EU states. This is of course nothing that can be forced upon Member States. However, I believe it is an absolute necessity if the EU is to live up to its ideals. (Malström, 2014c)

The assembled host is here turning on itself, disaggregating that 'self' and labelling its disgraceful elements based on an aspirational interpretation of its collective ethos. These criticisms have since been echoed by other Commissioners (e.g. Timmermans et al., 2015) and President Juncker (2015b), who used his 'State of the Union' speech in September 2015 to declare that 'There is not enough Europe in this Union. And there is not enough Union in this Union.' A minor victory was achieved later that month when the Commission's Relocation Programme (of 120,000 refugees) was approved by the Council of Ministers. Yet this was realised via a qualified majority, overriding the dissent of Hungary, Slovakia, the Czech Republic and Romania, undermining the well-established solidarity norm of consensus-based decision making (BBC, 2015b). Meanwhile, EUrope is still unable to agree a significant resettlement programme, let alone a hospitality and solidarity that would 'live up to its ideals'.

The differing understandings of hospitality and solidarity have also set member state against member state, most prominently Germany against Hungary. Having already accepted the largest number of asylum claims from Syrians, in 2015 Germany suspended the Dublin Regulation, a key element of the CEAS which allows countries to deport refugees back to the member state in which they first arrived. On 4 September, amidst the growing crisis in the Mediterranean and Western Balkans, Angela Merkel announced that she was opening Germany's borders to undocumented migrants, allowing some 3–7,000 people to arrive in Munich in one day to be met by food and cheering crowds (Graham-Harrison et al., 2015). Estimates for the number of asylum seekers Germany would receive in 2015 quickly rose from 800,000 towards 1.5 million. Such unilateral hospitality declared a solidarity with refugees but destroyed solidarity with fellow member states, particularly Hungary which effectively became a 'transit' state and whose Prime Minister, Viktor Orbán, strongly criticised German actions. Merkel hailed Germany's policy as offering a 'friendly, beautiful face' to the world (Harding, 2015) and used the widespread praise from civil society to pressure other member states into accepting quotas on relocation and resettlement. Such pressure, however, endangered not only solidarity, but also the democratic choices of states like Hungary where nearly 70 per cent of voters supported Orbán's hostility (Puhl, 2015).

German attempts to show solidarity with refugees appear to have failed, with border controls reimposed after eight days, causing confusion and anger amongst those en route to Germany (Kingsley, 2015). Meanwhile, Schengen states suspended free movement and Hungary erected a razor-wire fence along its border with Serbia to prevent further arrivals (Puhl, 2015). Hospitable solidarity was replaced by overt hostility. But due to its autoimmune nature, it is not clear which approach was more in line with EUrope's ethos. Germany's policy, for a brief period, more closely approached an unconditional hospitality that showed solidarity with, and respected the freedom and human rights of, those who require protection. It certainly appeared closer to EUropean ideals. Yet, in undermining EUrope's minimalist immunising hospitality, it effectively attacked EUrope's own immune system whilst destroying solidarity. Hungary's conservative hostility meanwhile was both democratic and broadly in line with EUrope's outsourced welcome. By providing 'no legal way to get to Europe' (Malström, 2014a), EUrope's autoimmunising hospitality has long forced migrants to cross the Mediterranean illegally, exposing refugees to death and 'threatening populations they are supposed to protect' (Vaughan-Williams, 2015: 116). Ultimately, the refugee crisis underlined the quasi-suicidal nature of EUrope's ethos, the ambivalences of which attack both its self and those to whom it offers protection.

## CONCLUSION

Ash Amin has argued that, as an increasingly multicultural Europe becomes a 'place of plural and strange belongings', the kind of values espoused as foundational to the EU have become 'a blunt instrument for unity' (2004: 2–3). Instead, Amin suggests embracing two principles, hospitality and mutuality, producing an 'imaginary of becoming European through engagement with the stranger' (ibid.: 4).

This, he argues, would be 'an inspiring and relevant ethos for a Europe distinguished by global ethnic and cultural mixture and intense mobility' (ibid.: 14). Amin is talking about a broader public ethos than the institutional ethos on which I have focused my attention. He is also concentrating on intra-European politics rather than EUrope's relation with its outside. Thus, he does not take account of the way EUrope's institutional ethos *already* constitutes and expresses itself through hospitality towards others.

This chapter has demonstrated that, like the other spaces examined in this book, EUrope is both produced through its practices of hospitality and exercises significant power in doing so. Yet where EUrope differs from other spaces is its innovative use of *protective* space, forming quasi-EUropean or (Trans-)EUropean spaces – the road to EUrope and the external dimension of asylum – into which strangers are welcomed as a means of sheltering both self and other from the risk they pose. Though the pastoral ethopolitics of enlargement and the pastoral biopolitics of immigration and asylum policy are very different, both share the same logic of caring for the stranger while immunising EUrope against threats from outside. Yet, because both emerge from an ambivalent and autoimmune ethos, EUrope cannot help but undermine its self *and* those it nominally welcomes and protects.

In the concluding chapter, I will outline how the dangers of autoimmunity are both an unavoidable risk and a necessary condition for any ethics and space of hospitality. While I would echo Amin's claim that an '[i]dea of Europe as hospitality towards the stranger' is inspiring, we cannot underplay its dangers, especially after they were starkly highlighted by the Paris attacks of November 2015. When combined with the ongoing refugee crisis, this violence appears to have left the Schengen agreement on the brink of potentially indefinite suspension (Traynor, 2016), casting doubt upon the future of the CEAS and EUrope as a space of protection, freedom, security and justice. French Prime Minister Manuel Valls has even suggested that the 'very idea of Europe' is at risk (BBC, 2016). But it is not refugees or migrants that have caused this; rather, it is the contradictions within EUrope's ethos which have been exposed through its practices of (auto)immunising hostipitality.

## NOTES

1 I am using 'EUrope' and' EUropean' to refer to the space and ethos of the EU which identifies itself with Europe as a whole but whose limits are not spatially nor legally coterminous (see Clark and Jones, 2008; Bialasiewicz, 2011; Bialasiewicz et al., 2012; Vaughan-Williams, 2015).

2 While all the migrants may not qualify as 'refugees' under the legal definition, I am following Al Jazeera's editorial policy of no longer talking about a 'migrant crisis' because the term is 'no longer fit for purpose when it comes to describing the horror unfolding in the Mediterranean' (Malone, 2015).

3 In ESPON's 2013 Programme ('Making Europe Open and Polycentric': Vision and Scenarios for the European Territory towards 2050), the preliminaries note the ambiguity of the 'space' whose planning they are monitoring – 'In the publication, "Europe" is associated to the ESPON space of 31 countries except when discussing common European policies, then "Europe" is associated to the European Union' (ESPON, 2013: iii).
4 Article 50 of the revised Treaty on European Union, which could shortly be triggered for the first time by the UK after the referendum 'Brexit' vote of 23 June 2016.
5 Cyprus, the Czech Republic, Estonia, Hungary, Latvia, Lithuania, Malta, Poland, Slovakia and Slovenia.
6 The CEE countries that gained accession in the 'big bang' were subject to a transition period of seven years before their citizens could work freely throughout EUrope. Similarly, entrants agreed to implement EUrope's border policies before membership without a commitment that existing member states would remove their own border controls in relation to the new members upon their accession (see Grabbe, 2006).
7 Through their membership of the European Economic Area (Norway) or bilateral agreements with the EU (Switzerland), they have access to the internal market, comply with most of its regulations and are members of Schengen, making them limners, neither fully inside nor outside the EUropean home (Kux and Sverdrup, 2000).
8 This term originates from Heather Grabbe (2006, 2014), senior adviser to Olli Rehn from 2004–9.
9 António Vitorino was Commissioner for Justice and Home Affairs from 1999 to 2004.
10 To this end, a number of Directives and Regulations were successfully passed and subsequently revised on the basis of their evaluation: the Asylum Procedures Directive, the Reception Conditions Directive, the Qualification Directive, the Eurodac Regulation, all of which have subsequently been revised alongside the Dublin Regulation (for an accessible summary, see European Commission, 2014).
11 Though the land border between Turkey and EUrope was effectively closed, this channelled refugees into the dangerous sea crossing to Greece, with nearly 44,000 using this route, a 272 per cent rise on 2013 (European Commission, 2015c: 2).
12 The much-publicised route through the Western Balkans saw only 20,000 crossings in 2013, increasing to over 200,000 by September 2015. See Frontex, 'Migratory Routes Map', http://frontex.europa.eu/trends-and-routes/migratory-routes-map/ (accessed 29 November 2015).
13 This seemed particularly hypocritical given that Ashton's predecessor Solana had overseen the brief existence and disintegration of 'Serbia and Montenegro' after the Montenegrins voted for secession (Friis, 2007).

# Conclusion: Risking Critical Practices of Hospitality

At 9.20pm on the 13 November 2015, EUrope's darkest fears about (auto) immunising hospitality appeared to be confirmed. With the German and French national football teams playing a friendly match at the Stade de France, a suicide-bomber blew himself up outside, killing one bystander. This began a set of suicide and gun attacks in the 10th and 11th arrondissements of Paris, culminating in three men entering the Bataclan concert hall at 9.40pm and firing for around 15 minutes into the audience of a rock concert. A siege commenced with concert-goers held as hostages for two hours before police stormed the hall at 12.20am. In total, 89 people were killed and over 200 were injured.[1] As the identities of the assailants became known in the following days, media attention focused on one man in particular, 'Ahmad al Mohammad', one of the suicide-bombers. Initially identified by the Syrian passport he was carrying, later found to be fake, 'al Mohammad's' fingerprints were confirmed by Eurodac to match those of a Syrian refugee who had arrived on the Greek island of Leros from Turkey on the 3 October 2015 (Lichfield, 2015). From there, it seems 'al Mohammad' took the Western Balkan route, moving from Greece to Macedonia, into Serbia, and from there into Croatia and Austria (Diehl and Reimann, 2015). The next time he was identified was via his remains in Paris (Faiola, 2015).

'Ahmad al Mohammad' and the ambiguity surrounding his identity encapsulated the dangers and risks of welcoming refugees and migrants – as the former chief of French intelligence observed, 'it is obvious now ... Amongst the migrants, there are some terrorists' (Faiola, 2015). Parties of the far-right across Europe were quick to seize on the 'refugee' involvement in the attacks, calling for the closing of national borders in contravention of Schengen agreements and for no more refugees to be welcomed. This was accompanied by a rapid rise in public support for these parties (Robins-Early, 2015), with France narrowly escaping a victory for the Le Front Nationale in regional elections. Hospitophobia was not confined to Europe; indeed, it was more mainstream in the US. 30 state governors declared that Syrian refugees were not welcome in their states, despite governors having no authority over the federal resettlement programme (Robbins, 2015). Hospitality rose to the top of the US presidential election agenda, with Republican candidates seeking to outdo each other in their hostility and Donald Trump suggesting that all Muslims be put under surveillance (Yuhas, 2015). On 7 December – three days after Tashfeen Malik and Syed Rizwan Farook had shot and killed 14 people, injuring 21 others in San Bernardinio, California (Schmidt and Pérez-Pēna, 2015) – Trump called for a 'total and complete shutdown of Muslims entering the United States' (Colvin, 2015).

## RISK AND AUTOIMMUNITY

Much of this hostility was initially prompted by one man, whose real identity may never be known. Doubts were cast on parts of 'al Mohammad's' journey when another man was arrested in Serbia with a copy of the same passport (*Guardian*, 2015b). Whether he was a refugee at all is unknown: he could easily have been a European Islamist known to the security services; his travelling as a refugee certainly succeeded in its possible intention, spreading fear through Europe (and the US) whilst discrediting Syrian refugees (Diehl and Reimann, 2015). Given that his fingerprints entered the Eurodac database in Greece, it seems certain that 'al Mohammad' made at least part of his journey as a refugee. He would therefore have benefited from the near unconditional hospitality of a postcolonial state, the humanitarian hospitality offered by at least one refugee camp (whether in Turkey, Greece, Macedonia or Serbia), the immunising hospitality of EUrope and the flourishing hospitality offered by Paris.

It is significant that the figure which embodies the worst fears of Western states is also that which unites all the post-sovereign spaces examined in this book. This is the irreducible risk posed by hospitality. As I noted in the Introduction, despite securitising governmental attempts to reduce the guest to one identity, this is not possible: the tourist may be a terrorist, the diplomat may be a spy, the refugee may be a radicalised Islamist. 'When I open my door for someone else, I open myself to someone who can destroy my home or my life, regardless of what rules I try to enforce on him or her or it' (Hagglund, 2008: 104). Here we have perhaps the biggest failure of hospitality's cinematic portrayal examined in Chapter 1: none of the films allow for the possibility that the other which is saved could be anything other than *worthy* a subject. They fail to account for the possibility that Tutsi 'refugees' in *Hotel Rwanda* could have been Hutu extremists in disguise; Raffi in *Ararat* could be, or become, a drug smuggler or a terrorist like his father; Emira in *Welcome to Sarajevo* could bring violence and unrest to the idealised feminine home. Furthermore, the right-wing reaction to 'al Mohammad' in both Europe and the US reveals the ultimate autoimmunity of hospitality in all the spaces examined: their openness to the stranger exposes them to their own destruction. But the opposite is also the case: closure destroys them *as spaces of hospitality*. This is the 'perversion and pervertibility' integral to hospitality, 'that one can become virtually xenophobic in order to protect or claim to protect one's own hospitality, the own home that makes possible one's own hospitality' (Derrida, 2000: 53).

If the risks of a necessarily autoimmune hospitality are irreducible, xenophobic attempts to sure up and protect the home are always already doomed to fail. The focus on 'al Mohammad' has shifted attention from the fact that, of the seven attackers identified by police, six were French nationals and EU citizens. The 'mastermind' of the attacks, Abdelhamid Abbaoud, was a Belgian national, born and raised in a Brussels suburb. While all of the known aggressors, bar the Abdeslam brothers, travelled to Syria in the past three years and fought for Islamic State, none was a recent immigrant. The national home that right-wing politicians were seeking to secure already contained its own insecurity, its non-belonging and non-self-identity, within itself. Banning migrants and refugees cannot change the fact that the 'sovereign' state has never managed to secure a monopoly on its own citizens' identification, belonging and sense of being-at-home. Indeed if, as noted in Chapter 4, postcolonial states are marked by the absence of internal pacification and a binding sense of 'we-ness' (Sørensen, 1997: 264), all states are postcolonial states.

This is why the Paris attacks are peculiarly relevant to this book. They were not easily interpreted as an attack *on* a nation or state, nor were they carried out *by* a nation or state. The targets were not the symbols of French government: the Élysée Palace or the French armed forces. Nor were they national cultural symbols – the Eiffel Tower or the Arc de Triomphe. They were the bars, restaurants, cafés, a laundrette, sports stadium and theatre in a mixed neighbourhood of a global city. In a thoughtful article, *Der Spiegel* draws attention to the significance of this:

> The 10th arrondissement between Gare de l'Est train station and the Canal Saint-Martin, and the 11th arrondissement between Place de la République and the Bastille, seem at first glance to be unlikely targets for a terrorist attack ... They are neighbourhoods where cafés and bars stand side-by-side, along with small green grocers, boutiques, Morrocan kebab houses, Chinese backroom eateries and halal butchers. Immigrants from Algeria, Tunisia, Sudan and Mali live here alongside university students from around the world, tradesmen, media professionals and aging anarchists. The religion of those who live here plays no role. The Occident and the Orient are crammed together here in a maze of small streets and make the best of it – a distillation of colorful, congenial, chaotic Europe at large. (*Spiegel* Staff, 2015)

The problematic celebration and *effacement* of difference in this picture of cosmopolitan mixity should not detract from its analysis. The effect, if not perhaps the intention which cannot be known, was a targeting of the openness, freedom and flourishing of everyday life created by the hospitality of Paris as a global city. These were spaces in which the attackers themselves – French/

Belgian nationals of North African descent – appeared to 'belong' as much as anyone else. The true horror of the attacks was that the liberty, anonymity and vulnerability of the global city was exposed and turned against its self. These events are not easily interpreted through the state-centric lens of IR; they are more easily read as a post-sovereign attack on a post-sovereign space. The attempt to refocus attention on Syria as the source of violence 'serves the purpose of disguising the fact that the attackers were ultimately locals' (*Spiegel* Staff, 2015). But it also draws attention to the fact that the wealth and vibrancy of the Parisian home is built on the past exploitation of other territories (postcolonial spaces) and the mobility of their previous inhabitants. While this history of interaction, responsibility and hospitality is denied by portraying the attack as an unprovoked 'act of war', the targeted neighbourhoods of Paris tell a more complex story.

But what can the arguments of this book offer in terms of a 'better' response or hospitality? What can a focus on hospitality as an ethics of post-sovereignty reveal that conventional statist approaches to international ethics – with their recourse to just wars, humanitarian interventions, global justice and foreign aid – miss? It has not been the aim of this book to provide policy suggestions or offer ethical guidance. However, given the horror of the attacks in Paris, the context in which I write this concluding chapter, it is necessary to revisit the arguments made whilst asking what an interpretation of the ethics and power relations involved in post-sovereign spaces of international hospitality add to a reading of these events.

## HOSPITALITY AS AN ETHICS OF POST-SOVEREIGNTY

The aim of this book was that of contributing to a movement beyond the state-oriented spatial imaginary of IR which largely restricts the way we conceive international ethics to a sovereign state agent bestowing generosity and just punishment outside its borders. The movement towards an ethics of post-sovereignty is one of not assuming state-centrality, but examining the 'habitus of experience', where ethical practises are part of everyday, ongoing encounters, and its subjects are 'unfinished, ambiguously located, and enigmatic', forever entangled in 'spatial practices' (Campbell and Shapiro, 1999b: xii–xvii). It therefore does not start with the need for normative judgement – what *ought* to be done in a given context – but begins from the context itself, the daily experience of international coexistence and how it is already being negotiated in the formation of an ethos, a way of being-in-relation to oneself and others. Hospitality is particularly significant to such an understanding of ethics as it defines this relation – it is a spatial practice which articulates an ethos in setting the affective terms of belonging and non-belonging, as well as the tactics of welcome and exclusion, freedom and control.

Thus, instead of asking 'how can France respond ethically to the Paris attacks', an ethics of post-sovereignty asks a range of 'how' questions, aiming to reveal the complexity, contingency and assumptions overlooked by normative international ethics. Questions such as: How are the space and ethos of Paris constantly being re-formed through the myriad ways it welcomes the stranger, controls them and secures itself against them? How is Paris, as the non-sovereign 'habitus of experience', *already* a product of ethical encounters, those that occur and those that remain potential? How is Paris always already responding to the threat of violence that resides within its home, where 'Occident' and 'Orient' are not 'crammed together' but disaggregated and reaggregated into a diverse range of non-binary, hybrid forms of being and belonging? And how is the space and ethos of Paris already related through histories of power, violence and exploitation, often forgotten or written-over, to other spaces whether they be sovereign, post-sovereign or non-sovereign? It is only by discounting such questions that we can expect a satisfactory answer to the normative question emerging from a statist-imaginary – how can France respond ethically? – and remain surprised when the answer is unsatisfactory and contentious. But it is only by answering them that we reveal how the 10th and 11th arrondissements, Paris, France, EUrope and other spaces are *already* responding ethico-politically to the challenge posed by radical difference and non-belonging.

An ethics of post-sovereignty can be approached in a range of ways, but the first of this book's three central arguments has been that exploring practices of hospitality is particularly urgent when migration and forced displacement are at an all-time high. In this context, not only *should* hospitality be taken seriously in the study of international ethics, it already *is* an important way in which everyday international ethical relations are being practised. As shown in Chapter 1, despite international ethics' disregard, popular culture, particularly cinema, has frequently stressed the centrality of hospitality as a response to genocide and ethnic cleansing. To make the case for hospitality as an important everyday ethical relation, I argued that we should think of it as a spatial practice, involving a dynamic relation between the inside and outside of a multi-scalar home, including affective dimensions of belonging and non-belonging.

Each chapter of the book therefore drew out how different spaces practise varied types of hospitality, depending upon their ethos as a 'home' and the way they police affective relations of belonging and non-belonging. Thus, Chapter 2 explored how refugee camps, especially in their idealised form, exhibit an ethos of temporary humanitarianism which regulates *how* they welcome and the limits of the care and attention they bestow on guests. Chapter 3 contrasted this with the ethos of freedom and flourishing displayed by global cities' targeting of, and turning a blind eye to, particular types of guests. In contrast to the caring control of the refugee camp, the global city's indifference to the welfare of its less fortunate guests is liberating, inhibiting and exploitative. This liberal indifference is central to understanding the global city's exposure and vulnerability to that which threatens it, both from within (the majority of the Paris attackers) and, perhaps, without ('al Mohammad'). Chapter 4 altered course, concentrating on a genealogy of how an ethos of near unconditional hospitality is foisted upon the postcolonial state, with fluctuating levels of coercion. In the case of (Trans-) Jordan, such extreme hospitality has become definitive of its international identity as a problematically 'sovereign' state. Finally, Chapter 5 exposed the (auto) immunising ethos of EUrope, which welcomes in a way that protects its 'home' and the values that define and 'found' it. The fact that these values ultimately undermine and contradict each other helps explain the current ethico-political crisis in which EUrope finds itself. These spaces illustrate some of the innumerable ways in which hospitality is part of the everyday practice of international ethical relations.

The second central argument of the book was that, not only do practices of hospitality generate subjects (hosts, guests and many shifting hybrids), they also actively produce spaces. On the one hand, none of these spaces would exist *as such* if it were not for their hospitality. The cinematic representations of hotel, homes and homelands were generated as places of idealised safety relative to their violent exterior, requiring an erasure or forgetting of the stranger's dangerous difference. The sole purpose of refugee camps as spaces is to welcome those most in need of protection and a temporary home. The global city could not be recognised as *global* without the cosmopolitan welcoming and circulation of different mobilities, including those that threaten its openness and existence. The postcolonial state as for ever pre-, post- or non-sovereign is rendered as such because it remains exceptionally open to aid, refugees, tourists and governance. And EUrope is constantly in the process of formation as a space because of the way it welcomes new members and outsources its values to spaces that become minimally 'EUropean'. On the other hand, these spaces are fashioned in their concrete particularity via distinct material and affective practices of hospitality. Building from Massey's (2005: 71) definition of space as a sphere of coexistence containing multiple trajectories moving into and out of contact with each other, I argued that hospitality is a key process by which a sphere is delimited and pacified, and the trajectories controlled. As such, hospitality works to produce *particular* spaces which are yours rather than mine, home rather than away, for circulation rather than stasis – for *this* rather than *that*.

It is here that the space-producing role of hospitality becomes entangled with my third argument regarding the inseparability of ethics and power relations. Far from a realm of purity and moral absolutes, ethical practices *require* the exercise of power in order to achieve, offer or *do* anything. For example, one cannot welcome a stranger into one's home without affirming sovereign mastery of the space – that it is *your* home – which automatically places limits on that welcome (Derrida, 2000: 14). Thus, even in cinema's attempt to portray hospitable acts of moral purity in the face of genocide, exercises of power are operating to muddy the waters. Whether this hospitality is enabled via sovereign power (*Hotel Rwanda*), patriarchal pastoralism (*Welcome to Sarajevo*) or securitising governmentalities (*Ararat*), welcome is conditioned on the silencing, erasure or forgetting of difference. Even apparently selfless hospitality produces the hotel, home or homeland through a modest cleansing of difference.

Chapter 2 concentrated on how the humanitarian hospitality of the idealised refugee camp operates through a form of domopolitics to generate a temporary home for the better protection and regulation of a suffering population. While this space does not reduce the refugee to bare life, its encouragement of a *minimal* homeliness, self-sufficiency and community amongst the displaced is also instrumental, aimed at better controlling their actions and emotions. While the freedoms offered by camps are often more apparent than substantive, they open possibilities for a counter-ethos which disrupts the humanitarian government of the camp space. Thus camps can become irruptive post-sovereign spaces in which national and cultural identities are created, entrenched or 'purified'. Meanwhile, domopolitical controls can never be assured because of the shifting identities and resistances of hosts and guests. It is no surprise that the camps 'al Mohammad' passed through en route to Paris could not contain or master the radical threat he posed.

In contrast, Chapter 3 drew out the way global cities seek to govern the trajectories and behaviour of their guests through tactics of security, managing host and guest alike via their freedom. The example of London illustrates how a space is created where everyone appears welcome, though some are targeted (the creative classes), some are excluded (the intolerant) and others are silently received via their exploitable transience (the (g)hosts). Mundane planning practices facilitate the creation of an urban zone which virtually belongs to the targeted guests and which (g)hosts pass through in performing the city's hosting work. As its flourishing ethos attempts to be largely self-governing, the security tactics of the global city can be countered by hosts (sanctuary movements) and guests (transience), though both counter-hospitalities reinforce as much as undermine its dominant mechanisms of governmentality. The Paris assailants performed a much more thorough-going form of counter-conduct which thoroughly revealed the autoimmunity of the global city's ethos. As both hosts and guests, they turned the freedoms of the global city against itself, using its indifference and liberalism to target and kill the very vibrant culture and way of life enabled by those security tactics. The resistant guest did not seek to become the host; rather, both sought to destroy the space and ethos of hospitality itself.

Unlike the hospitalities of camps and cities which emerge at the assembled hosts' instigation, the postcolonial state examined in Chapter 4 has hospitality thrust upon it. At first glance, the encounter between a weak host and a powerful guest opens this space to a near unconditional welcome. In the case of (Trans-) Jordan, a more complex genealogy explains how the 'sovereign' state was formed from the patriarchal, disciplinary pacification of a space, its hosts and its ethos. Guests who felt invited (the British and Hashemites) engaged in an elaborate destruction, reconstruction and correction of the borders, ungoverned wildernesses and culture of this space, leaving it a 'trans-' state, permanently *becoming*-sovereign. The openness which remains a hallmark of postcolonial (Trans-)Jordan means that its ethos can be further instrumentalised by later-arriving guests. These can be representatives of the international community (UNHCR, EUrope and humanitarian NGOs) who pay (Trans-)Jordan to take in refugees from Palestine, Iraq and Syria; or it can be USAID, which monetises and further disciplines (Trans-)Jordan's hospitality so it can be better sold to the international tourist market. The fact that postcolonial states like (Trans-)Jordan and Syria are only granted a form of sovereignty 'by courtesy' of Western states means that instability and civil war can always warrant further disciplinary interventions. Such actions can breed resentment and resistance, while the definitive openness of postcolonial spaces allows the easy circulation of revenge beyond their boundaries.

In this sense, spaces such as EUrope already contain their own destruction, despite attempts to immunise themselves through hospitality. The generation of EUropean space is hugely innovative, unique in international politics. Feeling insecure about its ephemeral boundaries and lack of a binding sense of belonging in the face of 'external' threats in the 1990s, EUrope constructed itself through a post-hoc 'founding' ethos which also defined its territoriality. As such, EUropean space could be exported and expanded, both through its ethopolitical enlargement and the biopolitical outsourcing of its 'protection' by raising asylum standards in other territories. This watchful, immunising hospitality would protect EUrope whilst caring for those it inclusively excluded. Yet this can never ultimately prove successful because its ethos and hospitality are *auto*immunising, leading it to attack its own principles of democracy and solidarity. EUrope has proven that it can never be identical with its self.

Furthermore, what the Paris attacks demonstrate is that the tolerance and multiculturalism EUrope and the global city celebrate as central to their ethos also undermines them. Their apparent outside already resides within. Tyranny, terror and intolerance inhabits their own history of exploiting the hospitality of now 'postcolonial' states. But it also dwells within the mixity that is constitutive of their spaces, whose homophobia, misogyny, racism and bigotry can never be fully policed. The democratic freedoms and liberties central to their way of life are precisely what *opens* them to attack from the inside. Thus 'al Mohammad' and his journey is incidental to events in Paris. As with Derrida's analysis of democracy's autoimmunity exemplified by 9/11, such spaces have two options: retain their ethos of hospitality and leave themselves open to attack from without and within; or cut back on their freedoms, fortify their borders, betray, corrupt and threaten their ethos in order to protect it (Derrida, 2005b: 40). Better management of their space, ethos and hospitality cannot resolve the contradictions internal to the values that define their belonging and non-belonging. Rather, all these spaces and their assembled hosts can do is seek to better understand and mitigate the worst effects of this autoimmunity, negotiating the space's security alongside its non-identical sense of self.

## 'EUROPEAN' SPACES OF PERPETUAL CRITIQUE

What, then, can we hope for from this analysis of hospitality as an ethics of post-sovereignty? Certainly we can expect no easy answers or ethical recommendations. In taking my lead from Derrida's reading of hospitality as the expression of an ethos, normative guidelines are already ruled out. While some have argued that Derrida's thought supports a radically unconditional hospitality, a space without borders and complete openness to whatever may come (e.g. Caputo, 2006: 278), this position is unsustainable. Hospitality *requires* a home with borders in order to operate, even as it constantly undermines, displaces and deconstructs it. Furthermore, even if it were possible to organise, an ethics of unconditional hospitality would be an automatic response, a renouncing of critical reflection, a welcoming of everything. This 'would short-circuit all forms of decisions and be the same as a complete indifference before what happens' (Hagglund, 2008: 103). Hospitality *must* discriminate and exclude; it must protect the home if it is to offer anything to host or guest. It must oppose that which would destroy its home, ethos, values and way of life, even while opening to that possibility. Home and hospitality are always a deconstructive negotiation and as such can never propose *an* ethics.

What a deconstructive ethics of post-sovereign hospitality is against, then, is the automatic, unreflective response to the arrival of the migrant or stranger. It is as much opposed to the campaign group 'No One Is Illegal' as to Donald Trump. One proposes the vision of a borderless, homeless world in opposing all controls on mobility and migration (No One Is Illegal Group, 2003); the other offers a xenophobically hostile homeland. What unites these seemingly radically opposed positions is that both destroy the conditions for spaces and practices of hospitality. One displays an automated indifference to whatever comes; the other offers a mechanical refusal of welcome. One rejects any exercise of power and so offers nothing; the other understands only sovereign exclusion. Neither offers hope to host or guest because they remove the conditions of possibility for such subjects. In place of both indeterminate and fortress spaces, and taking my lead from Derrida, I would provocatively and uncomfortably suggest that a deconstructive ethics of hospitality is searching for spaces which are becoming 'European'.

Shortly before his death in 2004, Derrida gave a brief speech to a Parisian audience celebrating the 50-year anniversary of *Le Monde diplomatique*. Entitled 'A Europe of Hope', Derrida begins with a keen awareness of the space in and from which he speaks – 'Paris, within France, speaking its language, and France within Europe' – asking whether the 'places that bear and take responsibility for this name' can take 'responsibility' for a different form of globalization (2006: 407). Here, he affirms a particular vision of 'Europe' which he previously referred to as the hospitable Europe of 'the other heading', whose culture is 'not to be identical to itself' (1992b: 9). This 'Europe of hope' would be one that 'sets the example of what a politics, a thinking, and an ethics could be, inherited from the passed Enlightenment and bearing the Enlightenment to come, which would be capable of non-binary judgements' (Derrida, 2006: 410). In opening towards the other, this 'Europe of Hope' would refer both to the 'historical thing called Europe', the continent, its institutions, but also to anyone and anywhere that affirms 'endless critical reflection' of its non-self-identical culture (Nass, 2008: 84–92). It could include refugee camps in the global south, global cities and postcolonial states everywhere, and EUrope wherever it may be. By affirming a passed Enlightenment, Derrida sees such European spaces as upholding many of the autoimmune values – democracy, freedom, equality, humanitarianism, justice and *hospitality* – which are central to the spaces examined in this book. It would also necessitate a 'responsible awareness' of Europe's 'guilty conscience', its 'totalitarian, genocidal, and colonialist crimes of the past' (Derrida, 2006: 410). But by referring to an 'Enlightenment to come' which is 'capable of non-binary judgements' Derrida affirms a relentlessly critical approach to Europe's history, values and future, such that it can never achieve a singular, purified, totalising vision of itself as a home, space and ethos.

This dream of 'European' spaces is far from singular and, while it shares a certain logic of values with EUropean space, would resist EUrope's current 'market-based' institutionalisation (Derrida, 2006: 410). Rather, it would be a Europe that is unremittingly critical and reflective, always rejecting easy answers and suspicious of claims to have acted or welcomed responsibly or ethically. It would resist satisfaction with the exclusionary autoimmunity of its ethos, regardless of the latter's necessity and irreducibility. It would always ask how it could welcome 'better', with less violence, whilst protecting its home against the risks it can never fully evade. It would wrestle with, and call itself to account for, past crimes of hospitality, colonialism and genocide. It would persistently question the power relations between hosts, guests and (g)hosts, the tactics and technologies of control that govern the space of the home – liberating, caring, securing, regulating, exploiting and resisting – as to whether they best express that space's problematic and contradictory ethos. It would constantly question the effects of its hospitality on other spaces, asking if more responsible relations are possible that would not drain resources without recompense.

But why label such perpetually critical and non-ideal spaces of hospitality 'European'? Obviously it opens whoever uses it to charges of Eurocentrism. After all, these values and critical self-reflection are thoroughly identified with the history of European Enlightenment (Nass, 2008: 84), along with its crimes. But it is in this sense that Europe remains 'irreplaceable' (Derrida, 2006: 410).

For Michael Naas, the label of 'Europe' is an example of 'paleonomy', the 'practice of reinscribing an old name in the name of a *promise* or even a *secret* harbored within that name' (2008: 84). But such an evasion of Eurocentrism is surely performed too easily, too comfortably. The fact remains that retaining and elevating 'Europe', a space identified with so many abuses of hospitality, ethics and power, is profoundly unsettling. But this is also its value as a label. 'European' spaces of hospitality gesture towards a hope for 'better' critical practices whilst never allowing this hope to *settle* into an ideal or regulative idea. Rather, they remain profoundly non-ideal, risky and discomfiting. Like hospitality.

## NOTE

1 For a fuller narrative, see *The Washington Post*'s (2015) and the *Guardian*'s (2015b) summaries. Unreferenced material below draws from here.

# References


Abdi, Awa M. (2005), 'In limbo: dependency, insecurity, and identity amongst Somali refugees in Dadaab camps', *Refuge: Canada's Journal on Refugees*, 22(2), 6–14.

Achilli, Luigi (2015), *Syrian Refugees in Jordan: A Reality Check*, Policy Brief, February (Firenze: Migration Policy Centre and European University Institute).

Adhikari, Mohamed (2008), '*Hotel Rwanda* – The challenges of historicising and commercialising genocide', *Development Dialogue*, 50(December), 173–195.

AFAD – Republic of Turkey Prime Ministry, Disaster and Emergency Management Presidency (2013), *Syrian Refugees in Turkey, 2013* (Ankara: AFAD).

Agamben, Giorgio (1995), 'We refugees', *Symposium*, 49(2), 114–119.

Agamben, Giorgio (1998), *Homo Sacer: Sovereign Power and Bare Life*, trans. Daniel Heller-Roazen (Palo Alto, CA: Stanford University Press).

Agier, Michel (2011), *Managing the Undesirables: Refugee Camps and Humanitarian Government*, trans. D. Fernbach (Cambridge: Polity Press).

Agnew, John (2007), 'Know-where: geographies of knowledge of world politics', *International Political Sociology*, 1(2), 138–148.

Ahmed, Sara (2000), *Strange Encounters: Embodied Others in Post-Coloniality* (London: Routledge).

Al Haija, Ahmed Abu (2011), 'Jordan: Tourism and conflict with local communities', *Habitat International*, 35(1), 93–100.

Alberti, Gabriella (2014), 'Mobility strategies, "mobility differentials" and "transnational exit": The experiences of precarious migrants in London's hospitality jobs', *Work, Employment and Society*, 28(6), 865–881.

Alon, Yoav (2009), *The Making of Jordan: Tribes, Colonialism and the Modern State* (London: Tauris).

Amin, Ash (2004), 'Multi-ethnicity and the idea of Europe', *Theory, Culture & Society*, 21(2), 1–24.

Amin, Ash (2012), *Land of Strangers* (Cambridge: Polity Press).

Amin, Ash and Nigel Thrift (2002), *Cities: Reimagining the Urban* (Cambridge: Polity Press).

Amstutz, Mark R. (2013), *International Ethics: Concepts, Theories, and Cases in Global Politics,* 4th edition (Lanham, MD: Rowman & Littlefield).

Anderson, Bridget (2010), 'Migration, immigration controls and the fashioning of precarious workers', *Work, Employment and Society*, 24(2), 300–317.

Andreas, Peter (2008), *Blue Helmets and Black Markets: The Business of Survival in the Siege of Sarajevo* (Ithaca, NY: Cornell University Press).

Andrijasevic, Rutvica (2010), 'From exception to excess: detention and deportations across the Mediterranean space', in Nicholas De Genova and Nathalie Peutz (eds),

*The Deportation Regime: Sovereignty, Space and the Freedom of Movement* (Durham, NC and London: Duke University Press), pp. 147–165.

Art, Robert J. and Kenneth Waltz (1983), 'Technology, strategy, and the uses of force', in Robert J. Art and Kenneth Waltz (eds), *The Use of Force*, 2nd edition (Lanham, MD: University of America Press), pp. 1–32.

Ashley, Richard K. (1987), 'The geopolitics of geopolitical space: towards a critical theory of international politics', *Alternatives*, 12(4), 402–432.

Ashton, Catherine (2010a), 'The EU and the Western Balkans in a changing world', Civil Society Meeting, Belgrade, 18 February.

Ashton, Catherine (2010b), Speech to the European Parliament on human rights, European Parliament, Strasbourg, 16 June.

Ashton, Catherine (2014), Speech at the National Assembly of the Republic of Serbia, Belgrade, 28 April.

Avramopoulos, Dimitri (2014a), European Security Forum 2014: Defining Europe's Priorities, European Security Roundtable, Brussels, 17 November.

Avramopoulos, Dimitri (2014b), European Commissioner Dimitris Avramopoulos calls in Geneva for more protection and admission possibilities for Syrian refugees, Geneva, 9 December.

Avramopoulos, Dimitri (2014c), Statement on the occasion of 'International Migrants Day', Brussels, 18 December.

Avramopoulos, Dimitri (2015), Opening remarks at Kos Press Conference, Kos, 4 September.

Bagelman, Jennifer (2013), 'Sanctuary: a politics of ease?', *Alternatives: Global, Local, Political*, 38(1), 49–62.

Baker, Gideon (2011), *Politicising Ethics in International Relations: Cosmopolitanism as Hospitality* (London: Routledge).

Banita, Georgiana (2012), '"The power to imagine": genocide, exile, and ethical memory in Atom Egoyan's *Ararat*', in Kristi M. Wilson and Tomás F. Crowder-Taraborrelli (eds), *Film and Genocide* (Madison, WI: University of Wisconsin Press), pp. 87–105.

Barnett, Clive (2005), 'Ways of relating: hospitality and the acknowledgement of otherness', *Progress in Human Geography*, 29(1), 5–21.

Barnett, Michael (2011), *Empire of Humanity: A History of Humanitarianism* (Ithaca, NY: Cornell University Press).

Barrot, Jacques (2008), 'The future of EU asylum policy: working towards a genuine area of protection', Ministerial conference 'Building a Europe of Asylum' extended to civil society, Paris, 8 September.

Batnitzky, Adina and Linda McDowell (2013), 'The emergence of an "ethnic economy"? The spatial relationships of migrant workers in London's health and hospitality sectors', *Ethnic and Racial Studies*, 36(12), 1997–2015.

Baumann, Zygmunt (2004), *Europe: An Unfinished Adventure* (Cambridge: Polity Press).

BBC (2013), 'Westgate attack: MPs to call for refugee camps to close', 30 September. Available at www.bbc.co.uk/news/world-africa-24339508 (accessed 4 March 2014).

BBC (2015a), 'Syrian refugee crisis in Jordan tests hospitality', 2 February. Available at www.bbc.co.uk/news/business-31058351 (accessed 16 November 2015).

BBC (2015b), 'Migrant crisis: Why EU deal on refugees is difficult', 25 September. Available at www.bbc.co.uk/newsworld-europe-34324096 (accessed 24 October 2015).

BBC (2015c), 'Migrant crisis: monthly record of 218,000 reach EU by sea', 2 November. Available at www.bbc.co.uk/news/world-europe-34700104 (accessed 6 November 2015).

BBC (2016), 'Migrant crisis: EU at grave risk, warns France PM Valls', 22 January. Available at www.bbc.co.uk/news/world-europe-35375303 (accessed 26 January 2016).

Bell, David (2007a), 'Moments of hospitality', in Jennie Germann Molz and Sarah Gibson (eds), *Mobilizing Hospitality: The Ethics of Social Relations in a Mobile World* (Aldershot: Ashgate), pp. 29–46.

Bell, David (2007b), 'The hospitable city: social relations in commercial spaces', *Progress in Human Geography*, 31(1), 7–22.

Bell, Duncan (ed.) (2010), *Ethics and World Politics* (Oxford: Oxford University Press).

Bell, Nathan (2014), 'France alone ...!', unpublished conference paper, *Hospitality Now (!)*, SciencesPo, Paris, 1 July.

Benhabib, Seyla (2004), *The Rights of Others: Aliens, Residents and Citizens* (Cambridge: Cambridge University Press).

Bennett, Bruce (2014), *The Cinema of Michael Winterbottom: Borders, Intimacy, Terror* (New York: Wallflower Press).

Benveniste, Emile (1973), *Indo-European Language and Society*, trans. E. Palmer (London: Faber).

Bialasiewicz, Luiza (2008), 'The uncertain state(s) of Europe', *European Urban and Regional Studies*, 15(1), 71–82.

Bialasiewicz, Luiza (ed.) (2011), *Europe in the World: EU Geopolitics and the Making of European Space* (Aldershot: Ashgate).

Bialasiewicz, Luiza (2012) 'Off-shoring and out-sourcing the borders of EUrope: Libya and EU border work in the Mediterranean', *Geopolitics*, 17(4), 843–866.

Bialasiewicz, Luiza, Stuart Elden and Joe Painter (2005), 'The constitution of EU territory', *Comparative European Politics*, 3(3), 333–363.

Bialasiewicz, Luiza, Paolo Giaccaria, Alun Jones and Claudio Minca (2012), 'Re-Scaling "EU"rope: EU macro-regional fantasies in the Mediterranean', *European Urban and Regional Studies*, 20(1), 59–76.

Bieber, Florian (2011), 'Building impossible states? State-building strategies and EU membership in the Western Balkans', *Europe-Asia Studies*, 63(10), 1783–1802.

Biedenkopf, Kurt, Bronislaw Geremek and Krzysztof Michalski (2004), 'The spiritual and cultural dimensions of Europe: concluding remarks', Reflection Group coordinated by the Institute for Human Sciences (Vienna/Brussels).

Biermann, Rafael (2014), 'Coercive Europeanization: the EU's struggle to contain secessionism in the Balkans', *European Security*, 23(4), 484–458.

Blanchard, Emmanuel, Bill MacKeith and Yasha Maccanico (2015), 'The EU and its neighbours: enforcing the politics of inhospitality', *Open Democracy*, 11 November. Available at www.opendemocracy.net/5050/emmanuel-blanchard/eu-forcing-politics-of-inhospitality-on-its-neighbours (accessed 1 December 2015).

Blunt, Alison and Robert Dowling (2006), *Home* (Abingdon: Routledge).

Bolchazy, Ladislaus J. (1993), *Hospitality in Antiquity: Livy's Concept of its Humanizing Force* (Chicago, IL: Ares).

Bradley, Megan (2013), 'Hospitality in crisis: reflections on Jordan's role in responding to Syrian displacement', *Brookings Up Front Blog*, 27 August. Available at www.brookings.edu/blogs/up-front/posts/2013/08/27-syrian-refugees-jordan-bradley (accessed 6 July 2015).

Bradshaw, Peter (2003), 'Ararat review', *Guardian*, 18 April. Available at www.theguardian.com/culture/2003/apr/18/artsfeatures8 (accessed 25 March 2014).

Brassett, James (2008), 'Cosmopolitanism vs. terrorism? Discourses of ethical possibility before and after 7/7', *Millennium: Journal of International Studies*, 36(2), 311–337.

Brickell, Katherine (2012), '"Mapping" and "doing" critical geographies of home', *Progress in Human Geography*, 36(2), 225–244.

Brown, Chris (2002), *Sovereignty, Rights and Justice: International Political Theory Today* (Cambridge: Polity Press).

Brown, Garrett Wallace (2010), 'The laws of hospitality, asylum seekers and cosmopolitan right: a Kantian response to Jacques Derrida', *European Journal of Political Theory*, 9(3), 308–327.

Bull, Hedley (1982), 'Civilian power Europe: a contradiction in terms?', *Journal of Common Market Studies*, 21(2), 149–164.

Bull, Hedley and Adam Watson (eds) (1984), *The Expansion of International Society* (Oxford: Oxford University Press).

Bulley, Dan (2006), 'Negotiating ethics: Campbell, ontopology and hospitality', *Review of International Studies*, 32(4), 645–663.

Bulley, Dan (2008), '"Foreign" terror? London bombings, resistance and the failing state', *British Journal of Politics and International Relations*, 10(3), 379–394.

Bulley, Dan (2009), *Ethics as Foreign Policy: Britain, the EU and the Other* (London: Routledge).

Bulley, Dan (2010), 'Home is where the human is? Ethics, intervention and hospitality in Kosovo', *Millennium: Journal of International Studies*, 39(1), 43–63.

Bulley, Dan (2013), 'Conducting strangers: hospitality and governmentality in the global city', in Gideon Baker (ed.), *Hospitality and World Politics* (Basingstoke: Palgrave Macmillan), pp. 222–245.

Bulley, Dan (2014a), 'Inside the tent: community and government in refugee camps', *Security Dialogue*, 45(1), 63–80.

Bulley, Dan (2014b), 'Foreign policy as ethics: towards a re-evaluation of values', *Foreign Policy Analysis*, 10(2), 165–180.

Bulley, Dan (2015), 'Ethics, power and space: international hospitality beyond Derrida', *Hospitality & Society*, 5(2–3), 185–201.

Bulley, Dan (2016), 'Occupy differently: space, community and urban counter-conduct', *Global Society*, 30(2), 238–257.

Bulley, Dan and Debbie Lisle (2012), 'Welcoming the world: governing hospitality in London's 2012 Olympic bid', *International Political Sociology*, 6(2), 186–204.

Burgess, John, Julia Connell and Jonathan Winterton (2013), 'Vulnerable workers, precarious work and the role of trade unions and HRM', *The International Journal of Human Resource Management*, 24(22), 4083–4093.

Butler, Judith (2004), *Precarious Life: The Powers of Mourning and Violence* (London and New York: Verso).

Camino, Mercedes Maroto (2005), '"The war is so young": masculinity and war correspondence in *Welcome to Sarajevo* and *Territorio Comanche*', *Studies in European Cinema*, 2(2), 115–124.

Campbell, David (1998), *National Deconstruction: Violence, Identity and Justice in Bosnia* (Minneapolis, MN: University of Minnesota Press).

Campbell, David and Michael J. Shapiro (1999b), 'Introduction: From ethical theory to the ethical relation', in David Campbell and Michael J. Shapiro (eds), *Moral Spaces: Rethinking Ethics in World Politics* (Minneapolis, MN: University of Minnesota Press), pp. vii–xx.

Candea, Matei (2012), '*Derrida en Corse?* Hospitality as scale-free abstraction', *Journal of the Royal Anthropological Institute*, 18(Issue supplement, s1): S34–48.
Candea, Matei and Giovanni Da Col (2012), 'The return to hospitality', *Journal of the Royal Anthropological Institute*, 18(Issue supplement, s1): S1–19.
Caney, Simon (2005), *Justice Beyond Borders: A Global Political Theory* (Cambridge: Cambridge University Press).
Caplan, Gerald (2009), 'Remembering Rwanda or denying it?', *Peace Review: A Journal of Social Justice*, 21(3), 280–285.
Caputo, John D. (2006), *The Weakness of God: A Theology of the Event* (Bloomington, IN: Indiana University Press).
Carens, Joseph H. (2003), 'Who should get in? The ethics of immigration admissions', *Ethics & International Affairs*, 17(1), 95–110.
Carrigan, Anthony (2011), *Postcolonial Tourism: Literature, Culture and the Environment* (London: Routledge).
Cavallar, Georg (2002), *The Rights of Strangers: Theories of International Hospitality, the Global Community and Political Justice since Vitoria* (Aldershot: Ashgate).
Chamayou, Grégoire (2012), *Manhunts: A Philosophical History*, trans. Steven Rendall (Princeton, NJ: Princeton University Press).
Christiansen, Thomas, Knud Erik Jorgensen and Antje Wiener (eds) (2001), *The Social Construction of Europe* (London: Sage).
Christopherson, Mona (2015), 'Jordan: Seeking Progress in the Land of Refugees', *International Peace Institute Global Observatory*, 3 February. Available online at: https://theglobalobservatory.org/2015/02/jordan-progress-land-refugees/ (accessed 15 May 2015).
City of Sanctuary (undated), 'Developing a culture of welcome', update to the *City of Sanctuary Handbook*. Available at http://cityofsanctuary.org/resources/background/culture/ (accessed 16 June 2015).
Clark, Julian and Alun Jones (2008), 'The spatialities of Europeanisation: territory, government and power in 'EUrope', *Transactions of the Institute of Political Geographers*, 33(3), 300–318.
Clegg, Nick (2014), 'UK to provide refuge to vulnerable Syrian refugees'. Available at www.gov.uk/government/news/uk-to-provide-refuge-to-vulnerable-syrian-refugees (accessed 1 February 2015).
Closs Stephens, Angharad (2007), '"Seven million Londoners, one London": national and urban ideas of community in the aftermath of the 7 July 2005 Bombings in London', *Alternatives: Global, Local, Political*, 32(2), 155–176.
Coaffee, Jon (2004), 'Rings of steel, rings of concrete and rings of confidence: designing out terrorism in Central London pre and post September 11th', *International Journal of Urban and Regional Research*, 28(1), 201–211.
Cochran, Molly (1999), *Normative Theory in International Relations: A Pragmatic Approach* (Cambridge: Cambridge University Press).
Colvin, Jill (2015), 'Trump calls for "complete shutdown" on Muslims entering US', *The Washington Post*, 7 December. Available at www.washingtonpost.com/politics/trump-calls-for-complete-shutdown-on-muslims-entering-us/2015/12/07/1f629a94-9d2a-11e5-9ad2-568d814bbf3b_story.html (accessed 14 December 2015).
Cook, Joanne, Peter Dwyer and Louise Waite (2011), 'The experiences of Accession 8 migrants in England: motivations, work and agency', *International Migration*, 49(2), 54–79.

Cooper, Robert (2002), 'The post-modern state', in Mark Leonard (ed.), *Re-ordering the World: The Long-term Implications of September 11* (London: Foreign Policy Centre), pp. 11–20.
Cooper, Robert (2003), *The Breaking of Nations: Order and Chaos in the Twenty-First Century* (London: Atlantic Books).
Corsellis, Tom and Antonella Vitale (executive editors and lead authors) (2005), *Transitional Settlement: Displaced Populations* (Cambridge: University of Cambridge shelterproject and Oxfam GB).
Council of the European Union (2009), 'Council Directive 2009/50/EC of 25 May 2009 on the conditions of entry and residence of third-country nationals for the purposes of highly qualified employment', *Official Journal of the European Union*, 18 June.
Coward, Martin (2012), 'Between us in the city: materiality, subjectivity, and community in the era of global urbanization', *Environment and Planning D: Society and Space*, 30(3), 468–481.
Craggs, Ruth (2014), 'Hospitality in geopolitics and the making of Commonwealth international relations', *Geoforum*, 52, 90–100.
Crisp, Jeff (2000), 'A state of insecurity: the political economy of violence in Kenya's refugee camps', *African Affairs*, 99(397), 601–632.
Crossrail (2014) 'Europe's largest construction project'. Available at www.crossrail.co.uk/construction/ (accessed 28 August 2014).
Darling, Jonathan (2009), 'Becoming bare life: asylum, hospitality and the politics of encampment', *Environment and Planning D: Society and Space*, 27(4), 649–665.
Darling, Jonathan (2010), 'A city of sanctuary: the relational re-imagining of Sheffield's asylum politics', *Transactions of the Institute of British Geographers*, 35(1), 125–140.
Darling, Jonathan (2011), 'Domopolitics, governmentality and the regulation of asylum accommodation', *Political Geography*, 30(5), 263–271.
Darling, Jonathan (2013), 'Moral urbanism, asylum, and the politics of critique', *Environment and Planning A*, 45(8), 1785–1801.
Davis, Mike (2006), *Planet of Slums* (London: Verso).
De Genova, Nicholas P. (2002), 'Migrant "illegality" and deportability in everyday life', *Annual Review of Anthropology*, 31, 419–447.
De Montclos, Marc-Antoine Perouse and Peter Mwangi Kagwanja (2000), 'Refugee camps or cities? The socio-economic dynamics of the Dadaab and Kakuma camps in Northern Kenya', *Journal of Refugee Studies*, 13(2), 205–222.
Death, Carl (2010), 'Counter-conducts: a Foucauldian analytics of protest', *Social Movement Studies*, 9(3), 235–251.
DeLanda, Manuel (2006), *A New Philosophy of Society: Assemblage Theory and Social Complexity* (London: Continuum).
Delanty, Gerard (1995), *Inventing Europe: Idea, Identity, Reality* (Basingstoke: Palgrave Macmillan).
Dennis, Darlene (2008), *Host or Hostage? A Guide for Surviving House Guests* (Encinitas, CA: Barthur House).
Derrida, Jacques (1992a), 'Force of law: the "mystical foundation of authority"', in Drucilla Cornell, Michael Rosenfeld and David Gray Carlson (eds), *Deconstruction and the Possibility of Justice* (London: Routledge), 3–67.
Derrida, Jacques (1992b), *The Other Heading: Reflections on Today's Europe*, trans. Pascale-Anne Brault and Michael B. Naas (Bloomington and Indianapolis, IN: Indiana University Press).

Derrida, Jacques (1998), *Monolingualism of the Other; or, The Prosthesis of Origin*, trans. P. Mensah (Palo Alto, CA: Stanford University Press).
Derrida, Jacques (1999a), *Adieu ... To Emmanuel Levinas*, trans. P.A. Brault and M. Naas (Palo Alto, CA: Stanford University Press).
Derrida, Jacques (1999b), 'Hospitality, justice and responsibility: a dialogue with Jacques Derrida', in Richard Kearney and Mark Dooley (eds), *Questioning Ethics: Contemporary Debates in Philosophy* (London: Routledge), pp. 65–83.
Derrida, Jacques (2000), 'Hostipitality', trans. B. Stocker and F. Morlock, *Angelaki: Journal of the Theoretical Humanities*, 5(3), 3–16.
Derrida, Jacques (2001), *On Cosmopolitanism and Forgiveness*, trans. Mark Dooley and Michael Hughes (London: Routledge).
Derrida, Jacques (2002a), *Acts of Religion* (New York: Routledge).
Derrida, Jacques (2002b), *Negotiations: Interventions and Interviews, 1971–2001*, trans. E. Rottenberg (Palo Alto, CA: Stanford University Press).
Derrida, Jacques (2003), 'Autoimmunity: real and symbolic suicides – a dialogue with Jacques Derrida', in Giovanna Borradori (ed.), *Philosophy in a Time of Terror: Dialogues with Jurgen Habermas and Jacques Derrida* (Chicago, IL: University of Chicago Press).
Derrida, Jacques (2005a), 'The principle of hospitality', *Parallax,* 11(1), 6–9.
Derrida, Jacques (2005b), *Rogues: Two Essays on Reason*, trans. Pascale-Anne Brault and Michael B. Naas (Palo Alto, CA: Stanford University Press).
Derrida, Jacques (2006), 'A Europe of hope', *Epoché: A Journal for the History of Philosophy*, 10(2), 407–412.
Derrida, Jacques and Anne Dufourmantelle (2000), *Of Hospitality*, trans. R. Bowlby (Palo Alto, CA: Stanford University Press).
Derrida, Jacques and Bernard Stiegler (2002), *Echographies of Television: Filmed Interviews*, trans. J. Bajorek (Cambridge: Polity Press).
Diehl, Jörg and Anna Reimann (2015), 'Paris attacks: how great are the terror dangers posed by refugees?', *Spiegel Online International*, 16 November. Available at www.spiegel.de/international/europe/paris-attacks-raise-questions-on-dangers-posed-by-refugees-a-1063026.html (accessed 14 December 2015).
Dikec, Mustafa (2002), 'Pera Peras Poros: longings for spaces of hospitality', *Theory, Culture & Society*, 19(1–2), 227–247.
Dikec, Mustafa, Nigel Clark and Clive Barnett (2009), 'Extending hospitality: giving space, taking time', *Paragraph*, 32(1), 1–14.
Diken, Bulent (2004), 'From refugee camps to gated communities – biopolitics and the end of the city', *Citizenship Studies*, 8(1), 83–106.
Diken, Bulent and Carsten Bagge Laustsen (2005), *The Culture of Exception: Sociology Facing the Camp* (London: Routledge).
Diken, Bulent and Carsten Bagge Laustsen (2006), 'The camp', *Geografiska Annaler. Series B, Human Geography*, 88(4), 443–452.
Dillon, Michael (1999), 'The scandal of the refugee: some reflections on the "inter" of international relations and continental thought', David Campbell and Michael J. Shapiro (eds), *Moral Spaces: Rethinking Ethics in World Politics* (Minneapolis, MN: University of Minnesota Press), 92–124.
Dillon, Michael (2007), 'Governing through contingency: the security of biopolitical governance', *Political Geography*, 26(1), 41–47.

Dinçer, Osman Bahadir, Vittoria Federici, Elizabeth Ferris, Sema Karaca, Kemal Kirişci and Elif Özmenek Çarmıklı (2013), *Turkey and Syrian Refugees: The Limits of Hospitality* (Washington, DC and Ankara: Brookings and USAK).

Dokotum, Okaka Opio (2013), 'Re-membering the Tutsi genocide in Hotel Rwanda (2004): implications for peace and reconciliation', *African Conflict & Peacebuilding Review*, 3(2), 129–150.

Donnelly, Jack (2000), *Realism and International Relations* (Cambridge: Cambridge University Press).

Doty, Roxanne Lynn (1996), *Imperial Encounters: The Politics of Representation in North–South Relations* (Minneapolis, MN: University of Minnesota Press).

Doty, Roxanne Lynn (2003), *Anti-Immigrantism in Western Democracies: Statecraft, Desire and the Politics of Exclusion* (London: Routledge).

Doty, Roxanne Lynn (2006), 'Fronteras Compasivas and the ethics of unconditional hospitality', *Millennium: Journal of International Studies*, 35(1), 53–74.

Duchêne, François (1972), 'Europe's role in world peace', in Richard Mayne (ed.), *Europe Tomorrow: Sixteen Europeans Look Ahead* (London: Fontana), pp. 32–47.

Duffield, Mark (2001), 'Governing the borderlands: decoding the power of aid', *Disasters*, 25(4), 308–320.

Duyvendak, Jan Willem (2011), *The Politics of Home: Belonging and Nostalgia in Western Europe and the United States* (Basingstoke: Palgrave Macmillan).

Ebert, Roger (2002), '*Ararat* review', *Chicago Sun-Times*, 22 November. Available at www.rogerebert.com/reviews/ararat-2002 (accessed 24 March 2014).

Edkins, Jenny (2003a), *Trauma and the Memory of Politics* (Cambridge: Cambridge University Press).

Edkins, Jenny (2003b), 'Humanitarianism, humanity, human', *Journal of Human Rights*, 2(2), 253–258.

Edkins, Jenny and Veronique Pin-Fat (2005), 'Through the wire: relations of power and relations of violence', *Millennium: Journal of International Studies*, 34(1), 1–24.

Ehata, Rebecca Kate (2013), *Migrant Belonging in International Relations: Tracing the Reflection of International Relations' Autochthonous Foundations in British Housing Discourse*, unpublished PhD thesis, University of Manchester.

Elden, Stewart (2007), 'Rethinking governmentality', *Political Geography*, 26(1), 29–33.

Enloe, Cynthia (1996), 'Margins, silences and bottom rungs: how to overcome the underestimation of power in the study of international relations', in Steve Smith, Ken Booth and Marysia Zalewski (eds), *International Theory: Positivism and Beyond* (Cambridge: Cambridge University Press), pp. 186–202.

Esposito, Roberto (2011), *Immunitas: The Protection and Negation of Life* (Cambridge: Polity Press).

European Commission (2002), Communication from the Commission to the Council and the European Parliament: Roadmaps for Bulgaria and Romania [COM(2002) 624 final], 13 November.

European Commission (2005), Communication from the Commission to the Council and the European Parliament on Regional Protection Programmes [COM(2005) 388 final], 1 September.

European Commission (2008a), 'Western Balkans: enhancing the european perspective', Communication from the Commission to the European Parliament and the Council [SEC(2008) 288], 5 March.

European Commission (2008b), 'Policy plan on asylum: an integrated approach to protection across the EU', Communication from the Commission to the European Parliament, the Council, the European Economic and Social Committee and the Committee of the Regions [SEC(2008) 2029/2030], 17 June.
European Commission (2011), 'The global approach to migration and mobility', Communication from the Commission to the European Parliament, the Council, the European Economic and Social Committee and the Committee of the Regions [COM(2015) 743 final /SEC(2011) 1353 final], 18 November.
European Commission (2013), 'Press Release: New EU regional development and protection programme for refugees and host communities in Lebanon, Jordan and Iraq', Brussels, 16 December.
European Commission (2014), *A Common European Asylum System* (Luxembourg: European Union Publications Office).
European Commission (2015a), 'A European agenda on migration', Communication from the Commission to the European Parliament, the Council, the European Economic and Social Committee and the Committee of the Regions [COM(2015) 240 final], 13 May.
European Commission (2015c), *The European Agenda on Migration: Glossary, Facts and Figures* (DG Migration and Home Affairs). Available at http://ec.europa.eu/dgs/home-affairs/e-library/multimedia/infographics/index_en.htm#080126248f65f02a/c (accessed 17 November 2015).
European Council (1993), Presidency Conclusions, Copenhagen European Council, 21–22 June.
European Council (1997), Presidency Conclusions, Luxembourg European Council, 12–13 September.
European Council (1999a), Presidency Conclusions, Tampere European Council, 15–16 October.
European Council (1999b), Presidency Conclusions, Helsinki European Council, 10–11 December.
European Council (2000), Presidency Conclusions, Santa Maria da Feira European Council, 19–20 June.
European Council (2002a), Presidency Conclusions, Seville European Council, 21–22 June.
European Council (2002b), Presidency Conclusions, Copenhagen European Council, 12–13 December.
European Council (2003), Presidency Conclusions, Brussels European Council, 12–13 December.
European Council (2004), Presidency Conclusions, Brussels European Council, 16–17 December.
European Council (2006), Presidency Conclusions, Brussels European Council, 14–15 December.
European Council (2009), Presidency Conclusions, Brussels European Council, 18–19 June.
European Council (2011), Conclusions, Brussels, 9 December.
European Council (2014), Conclusions, Brussels, 20–21 March.
European Council (2015a), Conclusions, Brussels, 19–20 March.
European Council (2015b), Conclusions, Brussels, 25–26 June.
European Council (2015c), Conclusions, Brussels, 17–18 December.
European Spatial Planning Observation Network (ESPON) (2013), *Making Europe Open and Polycentric: Visions and Scenarios for the European Territory towards 2050* (Luxembourg: ESPON 2013 Programme).

Evans, Yara, Jane Wills, Kavita Datta, Joanna Herbert, Cathy McIlwaine and Jon May (2007), '"Subcontracting by stealth" in London's hotels: impact and implications for labour reorganising', *Just Labour: A Canadian Journal of Work and Society*, 10(spring), 85–97.

Faiola, Anthony (2015), 'The mystery surrounding the Paris bomber with a fake Syrian passport', *The Washington Post*, 18 November. Available at www.washingtonpost.com/world/europe/the-mystery-surrounding-the-paris-bomber-with-a-fake-syrian-passport/2015/11/17/88adf3f4-8d53-11e5-934c-a369c80822c2_story.html (accessed 14 December 2015).

Fassin, Didier (2012), *Humanitarian Reason: A Moral History of the Present* (Berkeley, CA: University of California Press).

Florida, Richard (2002), *The Rise of the Creative Class ... and How it's Transforming Work, Leisure, Community & Everyday Life* (New York: Basic Books).

Florida, Richard (2005), *Cities and the Creative Class* (London: Routledge).

Forde, Chris and Robert MacKenzie (2009), 'Migrant workers in low-skilled employment: assessing the implications for human resource management', *International Journal of Manpower*, 30(5), 437–452.

Foucault, Michel (1982), 'The subject and power', *Critical Inquiry*, 8(4), 777–795.

Foucault, Michel (1991a), *Discipline and Punish: The Birth of the Prison*, trans. Alan Sheridan (London: Penguin).

Foucault, Michel (1991b), 'Nietzsche, genealogy, history', in Paul Rabinow (ed.), *The Foucault Reader: An Introduction to Foucault's Thought* (London: Penguin), pp. 77–100.

Foucault, Michel (1997), 'What is critique?', *The Politics of Truth* (New York: Semiotext[e]).

Foucault, Michel (1998), *The Will to Knowledge: The History of Sexuality Volume 1*, trans. Robert Hurley (London: Penguin).

Foucault, Michel (2004), *Society Must Be Defended: Lectures at the Collège de France, 1975–1976*, trans. David Macey (London: Penguin).

Foucault, Michel (2007), *Security, Territory, Population: Lectures at the Collège de France 1977–1978*, trans. Graham Burchell (New York: Picador).

Foucault, Michel (2008), *The Birth of Biopolitics: Lectures at the Collège de France 1978–1979*, trans. Graham Burchell (Basingstoke: Palgrave MacMillan).

Foundas, Scott (2004), 'Review: "Hotel Rwanda"', *Variety*, September 15. Available at http://variety.com/2004/film/reviews/hotel-rwanda-3-1200531037/ (accessed 14 February 2014).

Frattini, Franco (2005a), 'The Commission's policy priorities in the area of freedom, security and justice', Bundestag, Berlin, 14 February.

Frattini, Franco (2005b), 'The Hague Programme: our future investment in democratic stability and democratic security', Erasmus University of Rotterdam, Rotterdam, 23 June.

Frattini, Franco (2005c), 'Speaking points on "migration and asylum package"', press conference, Brussels, 1 September.

Frattini, Franco (2005d), 'Legal migration and the follow-up to the Green paper and on the fight against illegal immigration', University of Harvard, Boston, 7 November.

Frattini, Franco (2006a), 'Management of migration flows', joint debate – Freedom, Security and Justice – Immigration (EP), Strasbourg, 27 September.

Frattini, Franco (2006b), 'Chances and risks of migration and its significance for the security of the European Union', German Bundeskriminalamt Autumn Conference 2006, Wiesbaden, 16 November.

Frattini, Franco (2007a), 'The future of EU migration and integration policy', London School of Economics, London, 23 February.
Frattini, Franco (2007b), 'Enhanced mobility, vigorous integration strategy and zero tolerance on illegal employment: a dynamic approach to European immigration policies', High-level Conference on Legal Immigration, Lisbon, 13 September.
Friedman, John (1986), 'The world city hypothesis', *Development and Change*, 17(1), 69–84.
Friese, Heidrun (2004), 'Spaces of hospitality', trans. James Keye, *Angelaki: Journal of the Theoretical Humanities*, 9(2), 67–79.
Frieze, Donna-Lee (2008), 'Cycles of genocide, stories of denial: Atom Egoyan's *Ararat*', *Genocide Studies and Prevention*, 3(2), 243–262.
Friis, Karsten (2007), 'The referendum in Montenegro: the EU's postmodern diplomacy', *European Foreign Affairs Review*, 12, 67–88.
Füle, Štefan (2010a), Exchange of views on Iceland, Plenary Session, European Parliament, Strasbourg, 7 July.
Füle, Štefan (2010b), 'Serbia 10 years after: moving on towards the EU – what is expected of Serbia as a candidate country?', Public Hearing, European Parliament, Brussels, 16 September.
Füle, Štefan (2010c), Debate on EU enlargement in the Dutch Parliament, European Affairs Commission National Parliament, The Hague, 6 October.
Füle, Štefan (2010d), 'The EU: a force for peace, stability and prosperity in Wider Europe', Columbia University, New York, 30 November.
Füle, Štefan (2011a), 'Revolutionising the European Neighbourhood Policy in response to tougher Mediterranean revolutions', round table discussion organised by Members of the European Parliament, Brussels, 14 June.
Füle, Štefan (2011b), Statement at the EU–Croatia Intergovernmental Conference, EU–Croatia Intergovernmental Conference, Brussels, 30 June.
Füle, Štefan (2011c), 'EU accession of the Republic of Croatia', European Parliament Plenary Session, Brussels, 30 November.
Füle, Štefan (2012a), Opening speech at the launching of the High Level Accession Dialogue with Bosnia and Herzegovina, Brussels, 27 June.
Füle, Štefan (2012b), 'Ukraine and the world: addressing tomorrrow's challenges together', 9th Yalta Annual Meeting, Ukraine, 13 September.
Füle, Štefan (2013a), Address to the Parliamentary Assembly of the Council of Europe, Parliamentary Assembly of Council of Europe, Strasbourg, 24 January.
Füle, Štefan (2013b), 'Enlargement in perspective: how do candidate and potential candidate countries perceive accession in the light of the current crisis?', speech at Enlargement in Perspective Conference, Brussels, 6 March.
Füle, Štefan (2013c), 'What Bosnia and Herzegovina needs to do to move on EU path', Plenary session of the European Parliament, Brussels, 22 May.
Füle, Štefan (2013d), 'Enlargement: need for bold visions', Friends of Europe Conference: 'Western Balkans: fast lane, slow lane', 3 December.
Füle, Štefan (2014a), 'New Europe and enlargement in a new political context', speech at conference on 10 years of the Czech membership in the EU: 'The Czech Republic and Europe through each other's eyes', Prague, 11 April.
Füle, Štefan (2014b), 'Enlargement and Western Balkans: what's next?', Western Balkans Conference, Vienna, 3 June.

Gammeltoft-Hansen, Thomas (2011), 'Outsourcing asylum: the advent of protection lite', in Luiza Bialasiewicz (ed.), *Europe in the World: EU Geopolitics and the Making of European Space* (Aldershot: Ashgate), pp. 129–152.

Ganguly, Debjani (2007), '100 Days in Rwanda, 1994: trauma aesthetics and humanist ethics in an age of terror', *Humanities Research*, XIV(2), 49–65.

Geddes, Andrew (2014), 'The European Union', in James F. Hollifield, Philip L. Martin and Pia M. Orrenius (eds), *Controlling Immigration: A Global Perspective*, 3rd edition (Palo Alto, CA: Stanford University Press), pp. 433–451.

George, Rosemary Marangoly (1996), *The Politics of Home: Postcolonial Relations and Twentieth-Century Fiction* (Cambridge: Cambridge University Press).

Gibney, Matthew J. (2004), *The Ethics and Politics of Asylum: Liberal Democracy and the Response to Refugees* (Cambridge: Cambridge University Press).

Glover, Jonathan D. (2010), 'Genocide, human rights and the politics of memorialization: "Hotel Rwanda" and Africa's world war', *South Atlantic Review*, 75(2), 95–111.

Glubb, John Bagot (1948), *The Story of the Arab Legion* (London: Hodder & Stoughton).

Glubb, John Bagot (1957), *A Soldier with the Arabs* (London: Hodder & Stoughton).

Glubb, John Bagot (1971), *My Years with the Arabs* (Tunbridge Wells: Institute for Cultural Research).

Glubb, John Bagot (1983), *The Changing Scenes of Life: An Autobiography* (London: Quartet).

Golder, Ben (2007), 'Foucault and the genealogy of pastoral power', *Radical Philosophy Review*, 10(2), 157–176.

Gordon, Colin (1991), 'Governmental rationality: an introduction', in Graeme Burchell, Colin Gordon and Peter Miller (eds), *The Foucault Effect: Studies in Governmentality* (London: Harvester/Wheatsheaf), pp. 1–51.

Grabbe, Heather (2006), *The EU's Transformative Power: Europeanization through Conditionality in Central and Eastern Europe* (Basingstoke: Palgrave Macmillan).

Grabbe, Heather (2014), 'Six lessons of enlargement ten years on: the EU's transformative power in retrospect and prospect', in Nathaniel Copsey and Tim Haughton (eds), *The JCMS Annual Review of the European Union in 2013* (Chichester: Wiley), pp. 40–56.

Graham-Harrison, Emma, Patrick Kingsley, Rosie Waites and Tracy McVeigh (2015), 'Cheering German crowds greet refugees after long trek from Budapest to Munich', *Observer*, 5 September. Available at www.theguardian.com/world/2015/sep/05/refugee-crisis-warm-welcome-for-people-bussed-from-budapest (accessed 24 November 2015).

Greater London Authority (GLA) (2010), 'Who runs London?'. No longer available at www.london.gov.uk/who-runs-london (accessed 10 February 2010).

*Guardian* (2015a), 'Editorial – the *Guardian* view on the EU response to the refugee crisis: a challenge it has failed to meet', *Guardian*, 14 September. Available at www.theguardian.com/commentisfree/2015/sep/14/the-guardian-view-on-the-eu-response-to-the-refugee-crisis-a-challenge-it-has-failed-to-meet (accessed 24 October 2015).

*Guardian* (2015b), 'The men who attacked Paris: profile of a terror cell', *Guardian*, 9 December. Available at www.theguardian.com/world/ng-interactive/2015/nov/16/men-who-attacked-paris-profile-terror-cell (accessed 14 December 2015).

Gudehus, Christian, Steward Andersen and David Keller (2010), 'Understanding *Hotel Rwanda*: a reception study', *Memory Studies*, 3(4), 344–363.

Habermas, Jurgen (2009), *Europe: The Faltering Project* (Cambridge: Polity Press).

Hagglund, Martin (2008), *Radical Atheism: Derrida and the Time of Life* (Palo Alto, CA: Stanford University Press).
Hahn, Johannes (2015), 'European neighbourhood policy: the way forward', Vienna, 2 March.
Hanafi, Sara and Taylor Long (2010), 'Governance, governmentalities, and the state of exception in the Palestinian refugee camps in Lebanon', *Journal of Refugee Studies*, 23(2), 134–159.
Hansen, Peo (2009), 'Post-national Europe – without cosmopolitan guarantees', *Race & Class*, 50(4), 20–37.
Harding, Luke (2015), 'Angela Merkel defends Germany's handling of refugee influx', *Guardian*, 15 September. Available at www.theguardian.com/world/2015/sep/15/angela-merkel-defends-germanys-handling-of-refugee-influx (accessed 21 November 2015).
Hazbun, Waleed (2008), *Beaches, Ruins, Resorts: The Politics of Tourism in the Arab World* (Minneapolis, MN: University of Minnesota Press).
Heffernan, Michael J. (2000), *The Meaning of Europe: Geography and Geopolitics* (London: Arnold).
Herz, Manuel (ed.) (2013), *From Camp to City: Refugee Camps of the Western Sahara* (Zurich: Lars Muller).
Hoelzl, Michael (2004), 'Recognizing the sacrificial victim: the problem of solidarity for critical social theory', *Journal for Cultural and Religious Theory*, 6(1), 45–64.
Holden, Stephen (2002), 'To dwell on a historic tragedy or not: a bitter choice', *New York Times*, 15 November. Available at www.nytimes.com/movie/review?res=9903E6DA17 30F936A25752C1A9649C8B63 (accessed 25 March 2014).
Honig, Bonnie (1994), 'Difference, dilemmas, and the politics of the home', *Social Research*, 61(3), 563–597.
Horst, Cindy (2006), *Transnational Nomads: How Somalis Cope with Refugee Life in the Dadaab Camps of Kenya* (New York: Berghahn).
Hron, Madeleine (2012), 'Genres of "yet another genocide": cinematic representations of Rwanda', in Kristi M. Wilson and Tomás F. Crowder-Taraborrelli (eds), *Film and Genocide* (Madison, WI: University of Wisconsin Press), pp. 133–153.
Hubbard, Phil and Eleanor Wilkinson (2015), 'Welcoming the world? Hospitality, homonationalism, and the London 2012 Olympics', *Antipode: A Radical Journal of Geography*, 47(3), 598–615.
Human Rights Watch (2014), *World Report 2014* (New York: Seven Stories Press).
Human Rights Watch (2015), *World Report 2015* (New York: Seven Stories Press).
Hutchings, Kim (2007), 'Feminist ethics and political violence', *International Politics*, 44(1), 90–106.
Huysmans, Jeff (2006), *The Politics of Insecurity: Fear, Migration and Asylum in the EU* (Abingdon: Routledge).
Huysmans, Jeff (2008), 'The jargon of exception – on Scmitt, Agamben and the absence of political society', *International Political Sociology*, 2(2), 165–183.
Hyndman, Jennifer (2000), *Managing Displacement: Refugees and the Politics of Humanitarianism* (Minneapolis, MN: University of Minnesota Press).
Hyndman, Jennifer and Alison Mountz (2007), 'Refuge or refusal', in David Gregory and Allan Pred (eds), *Violent Geographies: Fear, Terror, and Political Violence* (London: Routledge), pp. 77–92.

Ilcan, Suzan and Anita Lacey (2011), *Governing the Poor: Exercises of Poverty Reduction, Practices of Aid* (London: McGill-Queen's University Press).

International Organization for Migration (IOM) (2015), *Mediterranean Update – Missing Migrants Project: 16 October 2015*. Available at http://missingmigrants.iom.int/sites/default/files/Mediterranean_Update_16_October.pdf (accessed 24 October 2015).

International Rescue Committee Commission on Syrian Refugees (2013), *Syria: A Regional Crisis* (New York: IRC).

Jabri, Vivienne (1998), 'Restyling the subject of responsibility in international relations', *Millennium: Journal of International Studies*, 27(3), 591–611.

Jabri, Vivienne (2010), 'Security, multiculturalism and the cosmopolis', in Angharad Closs Stephens and Nick Vaughan-Williams (eds), *Terrorism and the Politics of Response* (London: Routledge), pp. 44–59.

Jackson, Robert H. (1987), 'Quasi-states, dual regimes, and neoclassical theory: international jurisprudence and the third world', *International Organization*, 41(4), 519–549.

Jackson, Robert H. (1990), *Quasi-States: Sovereignty, International Relations and the Third World* (Cambridge: Cambridge University Press).

Jones, Alun (2011), 'Making regions for EU action: the EU and the Mediterranean', in Luiza Bialasiewicz (ed.), *Europe in the World: EU Geopolitics and the Making of European Space* (Aldershot: Ashgate), pp. 79–90.

Juncker, Jean-Claude (2015a), 'Tackling the migration crisis', speech at the debate in the European Parliament on the conclusions of the Special European Council on 23 April, Strasbourg, 29 April.

Juncker, Jean-Claude (2015b), State of the Union 2015: Time for Honesty, Unity and Solidarity, Strasbourg, 9 September.

Juncker, Jean-Claude (2015c), 'European Commission – Press Release: President Juncker launches the EU Emergency Trust Fund to tackle roots causes of irregular migration in Africa', Valletta, 12 November.

Juncos, Ana E. (2012), 'Member state-building versus peacebuilding: the contradictions of EU state-building in Bosnia and Herzegovina', *East European Politics*, 28(1), 58–75.

Kant, Immanuel (1991), *Political Writings*, 2nd edition, ed. H. Reiss (Cambridge: Polity Press).

Kaplan, Edward K. (2011), 'The open tent: angels and strangers', in Richard Kearney and James Taylor (eds), *Hosting the Stranger: Between Religions* (New York: Continuum), pp. 67–72.

Kearney, Richard and James Taylor (eds) (2011), *Hosting the Stranger: Between Religions* (New York: Continuum).

Khor, Lena (2011), 'Human rights and network power', *Human Rights Quarterly*, 33(1), 105–127.

Khosravi, Shahram (2010), *'Illegal' Traveller: An Auto-Ethnography of Borders* (Basingstoke: Palgrave Macmillan).

Kimmelman, Michael (2014), 'Refugee camp for Syrians in Jordan evolves as a do-it-yourself city', *New York Times*, 4 July. Available at www.nytimes.com/2014/07/05/world/middleeast/zaatari-refugee-camp-in-jordan-evolves-as-a-do-it-yourself-city.html?_r=0 (accessed 16 November 2015).

King Abdullah (I) Ibn Al Husayn (1950), *Memoirs of King Abdullah of Transjordan*, ed. Philip P. Graves, trans. G. Khuri (London: Jonathan Cape).

King Abdallah (I) ibn al-Husayn (1978), *My Memoirs Completed: 'Al Takmilah'*, trans. Harold W. Glidden (London: Longman).

King Abdullah II Ibn Al Hussein (2015), Remarks before the European Parliament, Strasbourg, 10 March. Available at http://kingabdullah.jo/index.php/en_US/speeches/view/id/552/videoDisplay/0.html (accessed 5 July 2015).

King Hussein (I) Ibn Talal (1978), 'A foreword', in King Abdallah (I) ibn al-Husayn, *My Memoirs Completed: 'Al Takmilah'*, trans. Harold W. Glidden (London: Longman Group), pp. v–xvii.

King Hussein I Ibn Al Talal (1962), *Uneasy Lies the Head: The Autobiography of His Majesty King Hussein I of the Hashemite Kingdom of Jordan* (New York: Bernard Geis).

King, Elisabeth (2010), 'Memory controversies in post-genocide Rwanda: implications for peace-building', *Genocide Studies and Prevention*, 5(3), 293–309.

Kingsley, Patrick (2015), 'Refugees confounded by Merkel's decision to close German borders', *Guardian*, 14 September. Available at www.theguardian.com/world/2015/sep/14/refugees-confounded-merkel-close-german-borders (accessed 1 October 2015).

Kinsella, Helen M. (2005), 'Securing the civilian: sex and gender in the laws of war', in Michael Barnett and Raymond Duvall (eds), *Power in Global Governance* (Cambridge: Cambridge University Press), pp. 249–272.

Kirişci, Kemal (2014), *Syrian Refugees and Turkey's Challenges: Going Beyond Hospitality* (Washington, DC: Brookings).

Knox, Dan (2011), *Tourism Cities* (London: Routledge).

Koonings, Kees and Dirk Krujit (eds) (2007), *Fractured Cities: Social Exclusion, Urban Violence & Contested Spaces in Latin America* (London: Zed Books).

Korf, Benedikt (2007), 'Antinomies of generosity: moral geographies of post-tsunami aid in Southeast Asia', *Geoforum*, 38(2), 366–378.

Kristeva, Julia (1991), *Strangers to Ourselves*, trans. Leon S. Roudiez (New York: Columbia University Press).

Kux, Stephan and Ulf Sverdrup (2000), 'Fuzzy borders and adaptive outsiders: Norway, Switzerland and the EU', *Journal of European Integration*, 22(3), 237–270.

Lahav, Gallya (2014), 'Commentary', in James F. Hollifield, Philip L. Martin and Pia M. Orrenius (eds), *Controlling Immigration: A Global Perspective*, 3rd edition (Palo Alto, CA: Stanford University Press), pp. 456–461.

Layne, Linda L. (1994), *Home and Homeland: The Dialogics of Tribal and National Identities in Jordan* (Princeton, NJ: Princeton University Press).

Lewis, Hannah, Peter Dwyer, Stuart Hodkinson and Louise Waite (2014), 'Hyper-precarious lives: migrants, work and forced labour in the Global North', *Progress in Human Geography*, OnlineFirst, 17 September.

Lichfield, John (2015), 'Paris terror: was there a 10th attacker and how many are still on the run?', *Independent*, 23 November.

Linklater, Andrew (1998), *The Transformation of Political Community* (Cambridge: Polity Press).

Lippert, Randy (1999), 'Governing refugees: The relevance of governmentality to understanding the International Refugee Regime', *Alternatives: Global, Local, Political*, 24(3), 295–328.

London Communications Agency (2011), 'Who runs London?' Available at www.lse.ac.uk/geographyAndEnvironment/research/london/pdf/WhoRunsLondon_2010.pdf (accessed 16 June 2015).

London Development Agency and Mayor of London (2009), *London Tourism Action Plan 2009–2013* (London: GLA).

London Development Agency and Mayor of London (2010), *The Mayor's Economic Development Strategy for London, May* (London: GLA).

London First (2015) 'Immigration'. Available at http://londonfirst.co.uk/our-focus/londons-workforce/immigration/ (accessed 16 June 2015).

Lucarelli, Sonia and Ian Manners (eds) (2006), *Values and Principles in European Union Foreign Policy* (London and New York: Routledge).

Lynch, Paul, Jennie Germann Molz, Alison McIntosh, Peter Lugosi and Conrad Lashley (2011), 'Theorizing hospitality', *Hospitality & Society*, 1(1), 3–24.

Magnusson, Warren (2011), *Politics of Urbanism: Seeing Like a City* (London: Routledge).

Malkki, Liisa H. (1995), *Purity and Exile: Violence, Memory and National Cosmology Among Hutu Refugees in Tanzania* (Chicago, IL: University of Chicago Press).

Malmberg, Bo, Eva Andersson and John Östh (2013), 'Segregation and urban unrest in Sweden', *Urban Geography*, 34(7), 1031–1046.

Malone, Barry (2015), 'Why Al Jazeera will not say Mediterranean "migrants"', Al Jazeera, 20 August. Available at www.aljazeera.com/blogs/editors-blog/2015/08/al-jazeera-mediterranean-migrants-150820082226309.html (accessed 24 October 2015).

Malström, Cecilia (2010a), 'Current challenges and opportunities in harmonising asylum and migration and the role of civil society', ECRE's (European Council on Refugees and Exiles) Presidents' and Directors Consultative Forum, Brussels, 24 June.

Malström, Cecilia (2010b), 'Establishing the Common European Asylum System by 2012 – an ambitious but feasible target', Ministerial Conference on Quality and Efficiency in the Asylum Process, Brussels, 14 September.

Malström, Cecilia (2011b), 'The European Asylum Support Office: implementing a more consistent and fair asylum policy', EASO Inaugural Event, Valletta, 19 June.

Malström, Cecilia (2011c), Statement of European Commissioner for Home Affairs, Cecilia Malmström, on the tragic loss of lives in the Mediterranean, Brussels, 5 August.

Malström, Cecilia (2011d), 'Higher standards of protection for refugees and asylum seekers in the EU', UNHCR Ministerial Meeting, Roundtable Discussion on Protection Challenges and Opportunities, Geneva, 7 December.

Malström, Cecilia (2012), 'Migration is an opportunity, not a threat', Global Hearing on Refugees and Migration, The Hague, 5 June.

Malström, Cecilia (2013a), 'EU Commission Press Release: Commissioner Malmström welcomes the European Parliament's vote on the Common European Asylum System', Strasbourg, 12 June.

Malström, Cecilia (2013b), Intervention during the press conference in Lampedusa, Lampedusa, 9 October.

Malström, Cecilia (2014a), 'An open and safe Europe – what's next?', Stakeholders Conference, Brussels, 29 January.

Malström, Cecilia (2014b), 'Common European Asylum System: challenges and perspectives', Conference on The Common European Asylum System: Challenges and Perspectives, Sofia, 24 March.

Malström, Cecilia (2014c), Statement – Commissioner Cecilia Malmström commemorates the Lampedusa tragedy, Brussels, 2 October.

Mancina, Peter (2013), 'The birth of a sanctuary-city: a history of governmental sanctuary in San Francisco', in Randy K. Lippert and Sean Rehaag (eds), *Sanctuary Practices in International Perspective: Migration, Citizenship and Social Movements* (Abingdon: Routledge), pp. 205–218.

Manners, Ian (2002), 'Normative power Europe: a contradiction in terms?', *Journal of Common Market Studies*, 40(2), 235–258.
Manners, Ian (2006), 'Normative power Europe reconsidered: beyond the crossroads', *Journal of European Public Policy,* 13(2), 182–199.
Manners, Ian (2008), 'The normative ethics of the European Union', *International Affairs*, 84(1), 45–60.
Marcuse, Peter and Ronald van Kempen (eds) (2000), *Globalizing Cities: A New Spatial Order?* (Oxford: Blackwell).
Marsden, Magnus (2012), 'Fatal embrace: trading in hospitality on the frontiers of South and Central Asia', *Journal of the Royal Anthropological Institute*, 18(Issue supplement, s1), 117–130.
Mason, Linda (2013), 'Jordan's hospitality for hundreds of thousands of Syrians', *MercyCorps*, 27 March. Available at www.mercycorps.org.uk/articles/jordan-syria/jordans-hospitality-hundreds-thousands-syrians (accessed 2 November 2015).
Massad, Joseph A. (2001), *Colonial Effects: The Making of National Identity in Jordan* (New York: Columbia University Press).
Massey, Doreen (1992), 'A place called home', *New Formations*, 17, 3–15.
Massey, Doreen (2004), 'Geographies of responsibility', *Geografiska Annaler*, 86B(1), 5–18.
Massey, Doreen (2005), *For Space* (London: Sage).
Massey, Doreen (2006), 'London inside-out', *Soundings*, 32, 62–71.
Massey, Doreen (2007), *World City* (Cambridge: Polity Press).
Matthews, Gareth and Martin Ruhs (2007), 'Are you being served? Employer demand for migrant labour in the UK's hospitality sector', Centre for Migration, Policy and Society, Working Paper No. 51, University of Oxford.
Mayor of London (2008), *Cultural Metropolis: The Mayor's Cultural Strategy – 2012 and Beyond* (London: Greater London Authority).
Mayor of London (2009), *The London Plan: Spatial Development Strategy for Greater London, consultation draft replacement* (London: Greater London Authority).
Mayor of London (2013), *2020 Vision – The Greatest City on Earth: Ambitions for London by Boris Johnson* (London: Greater London Authority).
Mayor of London (2014a), *Draft Further Alterations to The London Plan: The Spatial Development Strategy for Greater London*, January (London: Greater London Authority).
Mayor of London (2014b), *Cultural Metropolis: The Mayor's Cultural Strategy – Achievements and Next Steps* (London: Greater London Authority).
McClelland, Mac (2014), 'How to build a perfect refugee camp', *New York Times Magazine*, 13 February. Available at www.nytimes.com/2014/02/16/magazine/how-to-build-a-perfect-refugee-camp.html (accessed 4 March 2014).
McDowell, Linda, Adina Batnitzky and Sarah Dyer (2007), 'Division, segmentation, and interpellation: the embodied labors of migrant workers in a Greater London hotel', *Economic Geography*, 83(1), 1–25.
McNulty, Tracy (2007), *The Hostess: Hospitality, Femininity, and the Expropriation of Identity* (Minneapolis, MN: University of Minnesota Press).
Miller, Daniel (2008), *The Comfort of Things* (Cambridge: Polity Press).
Miller, Peter and Nikolas Rose (2008) *Governing the Present: Administering Economic, Social and Personal Life* (Cambridge: Polity Press).
Millner, Naomi (2011), 'From "refugee" to "migrant" in Calais solidarity activism: re-staging undocumented migration for a future politics of asylum', *Political Geography*, 30(6), 320–328.

Ministry of Tourism & Antiquities, Government of Jordan (MoTA) (2011), *Jordan: National Tourism Strategy 2011–2015* (Amman: Ministry of Tourism and Antiquities, Government of Jordan with the support of USAID).
Mitscherlich, Johanna (2013), 'Hospitality shines for Syrian refugees in Jordan', *CARE International blog*, 18 October. Available at www.care.org.au/blog/hospitality-shines-for-syrian-refugees-in-jordan/ (accessed 2 November 2015).
Molloy, Patricia (2000), 'Theatrical release: catharsis and spectacle in welcome to Sarajevo', *Alternatives: Global, Local, Political*, 25(1), 75–90.
Molz, Jennie Germann and Sarah Gibson (2007), 'Introduction: mobilizing and mooring hospitality', in Jennie Germann Molz and Sarah Gibson (eds), *Mobilizing Hospitality: The Ethics of Social Relations in a Mobile World* (Aldershot: Ashgate), pp. 1–26.
Morgenthau, Hans J. (1951), *In Defence of the National Interest: A Critical Examination of American Foreign Policy* (New York: Knopf).
Mulholland, Hélène (2009), 'Griffin: unfair that *Question Time* was filmed in "ethnically cleansed" London', *Guardian*, 23 October.
Mumford, Lewis (1938), *The Culture of Cities* (London: Secker & Warburg).
Murray, Graham (2006), 'France: the riots and the Republic', *Race & Class*, 47(4), 26–45.
Musliu, Vjosa (2014), 'A post-structuralist account of international missions: the case of Kosovo', unpublished PhD manuscript (University of Gent).
Mutua, Makau (2001), 'Savages, victims, and saviors: the metaphor of human rights', *Harvard International Law Journal*, 42(1), 201–245.
Nancy, Jean-Luc (1991), *The Inoperative Community* (Minneapolis, MN: University of Minnesota Press).
Nancy, Jean-Luc (2000), *Being Singular Plural* (Palo Alto, CA: Stanford University Press).
Nass, Michael (2008), *Derrida From Now On* (New York: Fordham University Press).
Ndahiro, Alfred and Privat Rutazindwa (2008), *Hotel Rwanda: Or the Tutsi Genocide as Seen by Hollywood* (Paris: L'Harmattan).
Neocleous, Mark (2003), *Imagining the State* (Maidenhead: Open University Press).
Neocleous, Mark (2011), '"A brighter and nicer new life": security as pacification', *Social & Legal Studies*, 20(2), 191–208.
Nicholson, Michael (1994), *Natasha's Story* (London: Pan/Macmillan).
No One Is Illegal Group (2003), *No One Is Illegal Manifesto*, 6 September. Available at www.noii.org.uk/no-one-is-illegal-manifesto/ (accessed 18 December 2015).
Noutcheva, Gergana (2009), 'Fake, partial and imposed compliance: the limits of the EU's normative power in the Western Balkans', *Journal of European Public Policy*, 16(7), 1065–1084.
Noutcheva, Gergana (2012), *European Foreign Policy and the Challenges of Balkan Accession: Conditionality, Legitimacy and Compliance* (Abingdon: Routledge).
Nye, Joseph S. (2004), *Soft Power: The Means to Success in World Politics* (New York: Public Affairs).
Nyers, Peter (2006), *Rethinking Refugees: Beyond States of Emergency* (London: Routledge).
Onuf, Nicholas Greenwood (1998), 'Everyday ethics in international relations', *Millennium: Journal of International Relations*, 27(3), 669–693.
Onuf, Nicholas (2013), 'Relative strangers: reflections on hospitality, social distance and diplomacy', in Gideon Baker (ed.), *Hospitality and World Politics* (Basingstoke: Palgrave Macmillan), pp. 173–196.
Orford, Anne (2003), *Reading Humanitarian Intervention: Human Rights and the Use of Force in International Law* (Cambridge: Cambridge University Press).

Orford, Anne (2007), 'Biopolitics and the tragic subject of human rights', in Elizabeth Dauphinee and Cristina Masters (eds), *The Logics of Biopower and the War on Terror: Living, Dying, Surviving* (New York: Palgrave Macmillan), pp. 205–228.

Osborne, Thomas and Nikolas Rose (1999), 'Governing cities: notes on the spatialisation of virtue', *Environment and Planning D: Society and Space*, 17(6), 737–760.

Owens, Patricia (2009), 'Reclaiming "bare life": against Agamben on refugees', *International Relations*, 23(4), 567–582.

Özden, Şenay (2013), 'Syrian refugees in Turkey', *Migration Policy Centre Research Reports 2013/05* (Robert Schuman Centre for Advanced Studies, San Domenico di Fiesole (FI): European Union Institute).

Papadopoulos, Dimitris, Niamh Stephenson and Vassilis Tsianos (2008), *Escape Routes: Control and Subversion in the Twenty-First Century* (London: Pluto Press).

Patsias, Caroline and Nastassia Williams (2013), 'Religious sanctuary in France and Canada', in Randy K. Lippert and Sean Rehaag (eds), *Sanctuary Practices in International Perspectives: Migration, Citizenship and Social Movements* (Abingdon: Routledge), pp. 175–188.

Patten, Chris (2000a), Speech to the Foreign Affairs and Legal Committees of the Albanian Parliament, Tirana, 6 March.

Patten, Chris (2000b), Speech to the Peace Implementation Council, Brussels, 23 May.

Patten, Chris (2000c), 'A common foreign policy for Europe: relations with Latin America', speech to the Consejo Argentino par alas Relaciones Internacionales (CARI), Buenos Aires, 9 November.

Patten, Chris (2001), 'EU strategy in the Balkans', The International Crisis Group, Brussels, 10 July.

Patten, Chris (2004), 'The Western Balkans: the road to Europe', speech to German Bundestag 'European Affairs Committee', Berlin, 28 April.

Patten, Chris (2005), *Not Quite the Diplomat: Home Truths About World Affairs* (London: Allen Lane).

Peck, Jamie (2005), 'Struggling with the creative class', *International Journal of Urban and Regional Research*, 29(4), 740–770.

Perera, Suvendrini (2002a), 'A line in the sea: The Tampa, boat stories and the border', *Cultural Studies Review*, 8(1), 11–27.

Perera, Suvendrini (2002b), 'What is a camp ...?', *borderlands e-journal*, 1(1).

Peteet, Julie (2005), *Landscape of Hope and Despair: Palestinian Refugee Camps* (Philadelphia, PA: University of Pennsylvania Press).

Peutz, Nathalie and Nicholas De Genova (2010), 'Introduction', in Nicholas De Genova and Nathalie Peutz (eds), *The Deportation Regime: Sovereignty, Space and Freedom of Movement* (Durham, NC and London: Duke University Press), pp. 1–30.

Philip, Abby (2015), 'A federal agent's gun was used in San Francisco "sanctuary city" murder case', *The Washington Post*, 8 July. Available at www.washingtonpost.com/news/morning-mix/wp/2015/07/08/san-francisco-murder-case-that-sparked-sanctuary-city-debate-takes-unexpected-turn/ (accessed 12 September 2015).

Phinnemore, David (2003), 'Stabilisation and association agreements: Europe agreements for the Western Balkans', *European Foreign Affairs Review*, 8(1), 77–103.

Phinnemore, David (2006), 'Beyond 25 – the changing face of EU enlargement: commitment, conditionality and the Constitutional Treaty', *Journal of Southern Europe and the Balkans*, 8(1), 7–26.

Phinnemore, David (2010), '"And We'd Like to Thank ...": Romania's integration into the European Union, 1989–2007', *Journal of European Integration*, 32(3), 291–308.

Pile, Steve (1999), 'The heterogeneity of cities', in Steve Pile, Christopher Brook and Gerry Mooney (eds), *Unruly Cities?* (Abingdon: Routledge), pp. 7–52.

Pippan, Christian (2004), 'The rocky road to Europe: the EU's stabilisation and association process for the Western Balkans and the principle of conditionality', *European Foreign Affairs Review*, 9(2), 219–245.

Pitt-Rivers, Julian, (2012), 'The law of hospitality', *HAU: Journal of Ethnographic Theory*, 2(1), 501–517 [Reprint of original book chapter from 1977].

Pogge, Thomas W. (1989), *Realizing Rawls* (Ithaca, NY: Cornell University Press).

Prodi, Romano (2000a), 'Europe and global governance', speech to 2nd COMECE Congress, Brussels, 31 March.

Prodi, Romano (2002b), 'The EU, the UK and the world', speech to Said Business School, Oxford, 29 April.

Prodi, Romano (2002c), 'Enlargement – the final lap', speech to European Parliament, Brussels, 9 October.

Prodi, Romano (2002d), 'The reality of enlargement', speech to the European Parliament, Brussels, 6 November.

Puhl, Jan (2015), 'Fortress Hungary: Orbán profits from the refugees', *Spiegel Online International*, 15 September. Available at www.spiegel.de/international/europe/viktor-orban-wants-to-keep-muslim-immigrants-out-of-hungary-a-1052568.html (accessed 1 October 2015).

Ramadan, Adam (2008), 'The guests' guests: Palestinian refugees, Lebanese civilians, and the war of 2006', *Antipode*, 40(4), 658–677.

Ramadan, Adam (2009), 'Destroying Nahr el-Bared: sovereignty and urbicide in the space of exception', *Political Geography*, 28(3), 153–163.

Ramadan, Adam (2010), 'In the ruins of Nahr el-Barid: understanding the meaning of the camp', *Journal of Palestine Studies*, 40(1), 49–62.

Ramadan, Adam (2013), 'Spatialising the refugee camp', *Transactions of the Institute of British Geographers*, 38(1), 65–77.

Refugee Council (2014), 'Refugee Council welcomes Syrian refugee commitment', 28 January. Available at www.refugeecouncil.org.uk/latest/news/3926_refugee_council_welcomes_syrian_refugee_commitment (accessed 1 February 2015).

Rehn, Olli (2004), 'Turkey and the EU: a common future?', Group meeting of the Greens/EFA of the European Parliament, Istanbul, 20 October.

Rehn, Olli (2005a), 'Values define Europe, not borders', speech to civil society, Belgrade, 24 January.

Rehn, Olli (2005b), 'Romania and the EU: common future, common challenges', speech at the Academy of Economic Studies, Bucharest, 28 February.

Rehn, Olli (2005c), 'Cyprus: one year after accession', Cyprus International Conference Center, Nicosia, 13 May.

Rehn, Olli (2005d), 'EU and Turkey together on the same journey', visit to Erciyes University, Kayseri, 7 October.

Rehn, Olli (2005e), 'Making the European Perspective real in the Balkans', Keynote address at the Bringing the Balkans into Mainstream Europe Conference, Friends of Europe, Brussels, 8 December.

Rehn, Olli (2006a), 'Enlargement in the evolution of the European Union', public lecture at the London School of Economics, London, 20 January.
Rehn, Olli (2006b), 'Beyond homogeneity', conference at the Central European University, Budapest, 9 February.
Rehn, Olli (2006c), 'Perspectives for Bosnia and Herzegovina', European Parliament, Strasbourg, 15 February.
Rehn, Olli (2006d), 'Building a new consensus on enlargement: how to match the strategic interest and functioning capacity of the EU?', European Policy Center, Brussels, 19 May.
Rehn, Olli (2006e), 'Turkey's best response is a rock-solid commitment to reforms', International Symposium on European Social Model and Trade Union Rights within the EU Negotiations, Ankara, 3 October.
Rehn, Olli (2006f), 'Europe's next frontiers', lecture at the Foreign Affairs Association, Munich, 20 October.
Rehn, Olli (2007), 'The European perspective for the Western Balkans', Western Balkans Panel at the International Conference Hall organised by Italian Ministry of Foreign Affairs, Rome, 16 January.
Rehn, Olli (2008a), 'Montenegro's journey to the EU', Parliament of Montenegro, Montenegro, 7 March.
Rehn, Olli (2008b), 'The EU strategy towards the Western Balkans and the role of Parliaments', Joint Parliamentary meeting at European Parliament, Brussels, 26 May.
Rehn, Olli (2008c), '2009: year of the Western Balkans and the Eastern Europe Partnership', public debate by DemosEuropa and Bertelsmann Foundation, Warsaw, 23 October.
Rehn, Olli (2008d), '45 years from the signing of the Ankara Agreement: EU-Turkey – cooperation continues', Conference on EC Turkey Association Agreement, 4 November.
Rehn, Olli (2009a), 'Five years of an enlarged EU', visit to Berlin, 28 April.
Rehn, Olli (2009b), 'Transforming the European Continent for the better', EU Enlargement 5 Years, Hässleholm, 11 May.
Rehn, Olli (2009c), 'Towards a European era for Bosnia and Herzegovina: the way ahead', Parliament of Bosnia and Herzegovina, Sarajevo, 24 July.
Rehn, Olli (2009d), 'EU Enlargement 2009: a balance sheet and way forward', Foreign Affairs Committee of the European Parliament, Brussels, 2 September.
Rehn, Olli (2009e), 'Enlargement and the EU's role in the world', University of Copenhagen, Copenhagen, 8 September.
Rehn, Olli (2009f), 'Enlargement package', European Parliament, AFET Committee, Brussels, 15 October.
Rehn, Olli (2009g), 'Lessons from EU enlargement for its future foreign policy', European Policy Centre, Brussels, 22 October.
Ridgley, Jennifer (2008), 'Cities of refuge: immigration enforcement, police, and the insurgent genealogies of citizenship in US sanctuary cities', *Urban Geography*, 29(1), 53–77.
Ridgley, Jennifer (2013), 'The city as a sanctuary in the United States', in Randy K. Lippert and Sean Rehaag (eds), *Sanctuary Practices in International Perspective: Migration, Citizenship and Social Movements* (Abingdon: Routledge), pp. 219–231.
Rieff, David (2002), *A Bed for the Night: Humanitarianism in Crisis* (London: Vintage).
Robbins, Liz (2015), 'Syrian family of 7 is resettled in New Jersey against Christie's opposition', *New York Times*, 30 November. Available at www.nytimes.com/2015/12/01/nyregion/syrian-family-of-7-is-quickly-settled-in-new-jersey.html (accessed 15 December 2015).

Robins-Early, Nick (2015), 'Europe's far right seeks to exploit post-Paris attack fears', *The Huffington Post*, 18 November. Available at www.huffingtonpost.com/entry/europe-far-right-paris-attacks_564b84bfe4b045bf3df16a03 (accessed 14 December 2015).

Roche, Maurice (2000), *Mega-events and Modernity: Olympics and Expos in the Growth of Global Culture* (Abingdon: Routledge).

Romney, Jonathan (2003), *Atom Egoyan* (London: BFI Publishing).

Rose, Nikolas (1999), *Powers of Freedom: Reframing Political Thought* (Cambridge: Cambridge University Press).

Rose, Nikolas (2000a), 'Community, citizenship, and the third way', *American Behavioral Scientist*, 43(9), 1395–1411.

Rose, Nikolas (2000b), 'Governing cities, governing citizens', in Engin F. Isin (ed.), *Democracy, Citizenship and the Global City* (London: Routledge), pp. 95–109.

Rose, Nikolas (2001), 'The politics of life itself', *Theory, Culture & Society*, 18(6), 1–30.

Rosello, Mireille (2001), *Postcolonial Hospitality: The Immigrant as Guest* (Palo Alto, CA: Stanford University Press).

Rosello, Mireille (2009), '"Wanted": organs, passports and the integrity of the transient's body', *Paragraph: Journal of Modern Critical Theory*, 32(1), 15–31.

Roth, Matt (1998), 'One-sided war', *Chicago Reader*, January 15. Available www.chicagoreader.com/chicago/one-sided-war/Content?oid=895342 (accessed 18 February 2014).

Rusesabagina, Paul with Tom Zoellner (2007), *An Ordinary Man: The True Story Behind Hotel Rwanda* (London: Bloomsbury).

Rygiel, Kim (2011), 'Bordering solidarities: migrant activism and the politics of movement and camps at Calais', *Citizenship Studies*, 15(1), 1–19.

Rygiel, Kim (2012), 'Politicizing camps: forging transgressive citizenship in and through transit', *Citizenship* Studies, 16(5–6), 807–825.

Samuel, Viscount (Herbert) (1945), *Memoirs* (London: Cresset Press).

Sanyal, Romola (2011), 'Squatting in camps: building and insurgency in spaces of refuge', *Urban Studies*, 48(5), 877–890.

Sanyal, Romola (2014), 'Urbanizing refuge: interrogating spaces of displacement', *International Journal of Urban and Regional Research*, 38(2), 558–572.

Sassen, Saskia (2001), *The Global City: New York, London, Tokyo*, 2nd edition (Princeton, NJ: Princeton University Press).

Schmidt, Michael and Richard Pérez-Pēna (2015), 'FBI treating San Bernardino attack as terrorism case', *New York Times*, 4 December. Available at www.nytimes.com/2015/12/05/us/tashfeen-malik-islamic-state.html?_r=1 (accessed 15 December 2015).

Selwyn, Tom (2000) 'An anthropology of hospitality', in Conrad Lashley and Alison Morrison (eds), *In Search of Hospitality: Theoretical Perspectives and Debates* (Oxford: Butterworth Heinemann), pp. 18–37.

Shapcott, Richard (2010), *International Ethics: A Critical Introduction* (Cambridge: Polity Press).

Shapiro, Michael J. (1994) 'Moral geographies and the ethics of post-sovereignty', *Public Culture*, 6, 479–502.

Shapiro, Michael J. (1997), 'Narrating the nation, unwelcoming the stranger: anti-immigration policy in contemporary America', *Alternatives: Global, Local, Political*, 22(1), 1–34.

Sharp, Jeremy M. (2010), *Jordan: Background and US Relations*, Congressional Research Service Report for Congress (Washington, DC: Congressional Research Service).

Shryock, Andrew (2004), 'The new Jordanian hospitality: house, host, and guest in the culture of public display', *Comparative Studies in Society and History*, 46(1), 35–62.

Shryock, Andrew (2008), 'Thinking about hospitality with Derrida, Kant, and the Balga Bedouin', *Anthropos*, 103(2), 405–421.

Shryock, Andrew (2009), 'Hospitality lessons: learning the shared language of Derrida and the Balga Bedouin', *Paragraph*, 32(1), 32–50.

Shryock, Andrew (2012), 'Breaking hospitality apart: bad hosts, bad guests, and the problem of sovereignty', *Journal of the Royal Anthropological Institute*, 18(Issue supplement, s1), S20–33.

Shryock, Andrew and Sally Howell (2001), '"Ever a guest in our house": the Emir Abdullah, Shaykh al-'Adwan, and the practice of Jordanian house politics, as remembered by Umm Sultan, the widow of Majid', *International Journal of Middle East Studies*, 33(2), 247–269.

Sicker, Martin (2001), *The Middle-East in the 21st Century* (Westport, CT: Praeger).

Sjoberg, Laura (2014), *Gender, War & Conflict* (Cambridge: Polity Press).

Smith, David M. (2000), *Moral Geographies: Ethics in a World of Difference* (Edinburgh: Edinburgh University Press).

Soguk, Nevzat (1999), *States and Strangers: Refugees and Displacements of Statecraft* (Minneapolis, MN: University of Minnesota Press).

Sokhi-Bulley, Bal (2015), 'Performing struggle: *Parrhēsia* in Ferguson', *Law and Critique*, 26(1), 7–10.

Solana, Javier (2000a), Speech to the Fernandez Ordonez Seminar, 14 January.

Solana, Javier (2000b), Speech to the European Parliament, 1 March.

Solana, Javier (2000c), 'The European Union is assisting the recovery – but much work remains to be done', *The Wall Street Journal*, 24 March.

Solana, Javier (2001), 'EU foreign policy', speech at the Hendrik Brugmans Memorial, Bruges, 25 April.

Solana, Javier (2003), Interview with *Dnevi Avaz*, *BiH* newspaper, 24 September.

Solana, Javier (2005), Speech at the Inauguration of the Academic Year, College of Europe, Bruges, 19 October.

Solana, Javier (2009a), 'Together we are stronger', University College Dublin, Dublin, 22 April.

Solana, Javier (2009b), 'Europe's global role – what next steps?', Ditchley Foundation Lecture, London, 11 July.

Sørensen, Georg (1997), 'An analysis of contemporary statehood: consequences for conflict and cooperation', *Review of International Studies*, 23(3), 253–269.

Spence, Lorna (2005), *Country of Birth and Labour Market Outcomes in London: An Analysis of Labour Force Survey and Census Data* (London: Greater London Authority).

Sphere Project (2011), *Humanitarian Charter and Minimum Standards in Humanitarian Response*, 3rd edition (Northampton: Practical Action).

*Spiegel* Staff (2015), 'The Paris attacks: murderous hatred in the city of love', *Spiegel Online International*, from *Der Spiegel*, 18 November. Available at www.spiegel.de/international/europe/paris-attacks-pose-challenge-to-european-security-a-1063435.html (accessed 14 December 2015).

Squire, Vicki and Jonathan Darling (2013), 'The "minor" politics of rightful presence: justice and relationality in the city of sanctuary', *International Political Sociology*, 7(1), 59–74.

Standing, Guy (2014), *The Precariat: The New Dangerous Class* (London: Bloomsbury Academic).
Steiner, George (2015), *The Idea of Europe: An Essay* (London: Overlook Duckworth).
Still, Judith (2007), 'Figures of Oriental hospitality: Nomads and Sybarites', in Jennie Germann Molz and Sarah Gibson (eds), *Mobilizing Hospitality: The Ethics of Social Relations in a Mobile World* (Aldershot: Ashgate), pp. 193–208.
Still, Judith (2010), *Derrida and Hospitality: Theory and Practice* (Edinburgh: Edinburgh University Press).
Still, Judith (2011), *Enlightenment Hospitality: Cannibals, Harems and Adoption* (Oxford: Voltaire Foundation).
Sylvester, Christine (1994), *Feminist Theory and International Relations in a Postmodern Era* (Cambridge: Cambridge University Press).
Taylor, Peter J. (2004), *World City Network: A Global Urban Analysis* (London: Routledge).
Thomas, Ward (2001), *The Ethics of Destruction: Norms and Force in International Relations* (Ithaca, NY: Cornell University Press).
Timmermans, Frans, Federica Mogherini, Neven Mimica, Dimitris Avramopoulos and Christos Stylianides (2015), Joint statement ahead of World Refugees Day on 20 June, Brussels, 19 June.
Tonkiss, Fran (2005), *Space, the City and Social Theory: Social Relations and Urban Forms* (Cambridge: Polity Press).
Traynor, Ian (2016), 'EU border controls: Schengen scheme on the brink after Amsterdam talks', *Guardian*, 26 January.
UNHCR (2007), *Handbook for Emergencies*, 3rd edition (Geneva: UNHCR).
UNHCR (2013), *Global Trends 2012: Displacement – The New 21st Century Challenge* (Geneva: UNHCR).
UNHCR (2014a), *2014 Syria Regional Response Plan: Jordan. Mid-Year Update* (Geneva: UNHCR).
UNHCR (2014b), *Global Trends 2013: War's Human Cost* (Geneva: UNHCR).
UNHCR (2015), *Global Trends 2014: World at War* (Geneva: UNHCR).
Uraizee, Joya (2010), 'Gazing at the beast: describing mass murder in Deepa Metha's *Earth* and Terry George's *Hotel Rwanda*', *Shofar: An Interdisciplinary Journal of Jewish Studies*, 29(4), 10–27.
USAID (2008), *Jordan Tourism Development Project: Final Report – October 2008* (Washington, DC: USAID).
USAID (2012), 'Where does USAID's money go?' Available at www.usaid.gov/documents/1870/where-does-money-go-excel-spreadsheet (accessed 13 November 2015).
USAID (2013), *Jordan Tourism Development Project II: Final Report – 2008–2013* (Washington, DC: USAID and Chemonics International).
Vaughan-Williams, Nick (2009), *Border Politics: The Limits of Sovereign Power* (Edinburgh: Edinburgh University Press).
Vaughan-Williams, Nick (2015), *Europe's Border Crisis: Biopolitical Security and Beyond* (Oxford: Oxford University Press).
Vitorino, Antonio (2001), 'The European dimension of immigration, policing and crime', Puglia Regional Council, Puglia, 19 June.
Walker, R.B.J. (1993), *Inside/Outside: International Relations as Political Theory* (Cambridge: Cambridge University Press).

Walters, William (2004), 'Secure borders, safe haven, domopolitics', *Citizenship Studies*, 8(3), 237–260.

Walters, William (2008), 'Acts of demonstration: mapping the territory of (non-) citizenship', in Engin F. Isin and Greg M. Nielsen (eds), *Acts of Citizenship* (London: Zed Books), pp. 182–206.

Walters, William (2011), 'Foucault and frontiers: notes on the birth of the humanitarian border', in Ulrich Bröckling, Susanne Krasmann and Thomas Lemke (eds), *Governmentality: Current Issues and Future Challenges* (New York: Routledge), pp. 138–164.

Walters, William (2012), *Governmentality: Critical Encounters* (London: Routeldge).

Waltz, Kenneth N. (1979), *Theory of International Politics* (Reading: Addison-Wesley).

Walzer, Michael (1973), 'Political action: the problem of dirty hands', *Philosophy & Public Affairs*, 2(2), 160–180.

Walzer, Michael (1983), *Spheres of Justice: A Defense of Pluralism and Equality* (New York: Basic Books).

Walzer, Michael (2006), *Just and Unjust Wars: A Moral Argument with Historical Illustrations*, revised edition (New York: Basic Books).

*Washington Post* (2015), 'What we know about the Paris attacks and the hunt for the attackers', *The Washington Post*, 9 December. Available at www.washingtonpost.com/graphics/world/paris-attacks/ (accessed 13 December 2015).

Weston, Phoebe (2015), 'Inside Zaatari refugee camp: the fourth largest city in Jordan', *Telegraph*, 5 August. Available at www.telegraph.co.uk/news/worldnews/middleeast/jordan/11782770/What-is-life-like-inside-the-largest-Syrian-refugee-camp-Zaatari-in-Jordan.html (accessed 16 November 2015).

Wheeler, Nick (1999), *Saving Strangers: Humanitarian Intervention in International Society* (Oxford: Oxford University Press).

Wilkinson, Mick (2012), 'Out of sight, out of mind: the exploitation of migrant workers in 21st Century Britain', *Journal of Poverty and Social Justice*, 20(1), 13–21.

Williams, Andrew (2010), *The Ethos of Europe: Values, Law and Justice in the EU* (Cambridge: Cambridge University Press).

Williams, John (2003), 'Territorial borders, international ethics and geography: do good fences still make good neighbours?', *Geopolitics*, 8(2), 25–46.

Wills, Jane (2012), 'The geography of community and political organisation in London today', *Political Geography*, 31(2), 114–126.

Wills, Jane, Kavita Datta, Yara Evans, Joanna Herbert, Jon May and Cathy McIlwaine (2010), *Global Cities at Work: New Migrant Divisions of Labour* (London: Pluto Press).

Wills, Jane, Jon May, Kavita Datta, Yara Evans, Joanna Herbert and Cathy McIlwaine (2009), 'London's migrant division of labour', *European Urban and Regional Studies*, 16(3), 257–271.

Wilson, Kristi M. and Tomás F. Crowder-Taraborrelli (eds) (2012), *Film and Genocide* (Madison, WI: University of Wisconsin Press).

Wilson, Mary C. (1987), *King Abdullah, Britain and the Making of Jordan* (Cambridge: Cambridge University Press).

Wirth, Louis (1938), 'Urbanism as a way of life', *American Journal of Sociology*, 44(1), 1–24.

World Bank (2015), 'Data: net official development assistance and official aid received (current US$)'. Available at http://data.worldbank.org/indicator/DT.ODA.ALLD.CD (accessed 7 July 2015).

Yamashita, Hikaru (2004), *Humanitarian Space and International Politics: The Creation of Safe Areas* (Aldershot: Ashgate).

Yuhas, Alan (2015), 'Trump won't rule out special ID for Muslim Americans noting their religion', *Guardian*, 20 November. Available at www.theguardian.com/us-news/2015/nov/19/donald-trump-muslim-americans-special-identification-tracking-mosques (accessed 15 December 2015).

Zielonka, Jan (2006), *Europe as Empire: The Nature of the Enlarged European Union* (Oxford: Oxford University Press).

# Index

*Note*: page numbers in italics denote figures